FEAST of Master Chef
by Alvin Liu

大廚 小宴

入廚新手學技巧，
在家宴客零難度！

序

賓客吃得稱心滿意，讀者在烹調上得到讚賞，是我編寫食譜的賞心樂事！

不經不覺，上一次出版食譜已是一年多前的事了，在這段期間，不時與友人飯聚聊天，他們在烹飪上遇到最惱人的問題是，在家宴客時如何選購材料、如何從容不迫地準備一席家庭宴。有見及此，我以多年教授烹飪的經驗，將解決方法集結成最新的食譜作品——《大廚小宴》，從設計菜單開始，為你介紹材料的特性及選購要點，並且預早一天恰當地處理食材，在飯宴當天不慌不忙輕鬆烹調，令客人吃得滿意！

此書籌備了一年時間，每部份都經過細心策劃，特別設計了六款不同主題的餐單，適合夏季、秋季、春節賀年及素食享用，每款餐單包含八個食譜，以 8 至 10 人份量為主（若人數不足，可將份量減半），賓客先品嘗前菜、湯羹，到主菜、飯麵及甜品等，吃得盡興之餘，也可感受你細緻的安排。

在這裏本人特別鳴謝林勤樂師傅及陳永瀚師傅給予此書寶貴的意見，更多謝我的廚務助手 Nicole，在拍攝期間的用心安排及幫助，以及感謝各方好友的支持，令食譜更加盡善盡美！

我期望此書為你帶來宴客的靈感，在烹調宴客菜時得心應手。

廖教賢

即時睇片 ┈┈┈▶

夏天饗客

前菜
酥脆響鈴　　14
Crisp Beancurd Skin Rolls

湯羹
魚肚豆腐羹　　16
Fish Maw and Tofu Thick Soup

主菜
白玉火鴨卷　　19
Winter Melon and Roast Duck Rolls

鮮露筍炒蝦片　　22
Stir-fried Asparagus and Shrimp Slices

香橙脆肉脯　　25
Deep-fried Pork in
Spiced Orange Sauce

香茅鹽焗烏頭魚　　28
Salt-baked Grey Mullet with
Lemongrass

飯麵
上湯水餃麵　　30
Dumplings and Noodles in Soup

甜品
綠茶凍糕　　34
Cold Green Tea Pudding

秋天宴會

前菜
麻辣蒜香鴨舌　　38
Spicy Garlic Duck Tongues

湯羹
砂窩雲吞雞　　41
Wontons and Chicken in Casserole

主菜
荷葉籠仔蒸糯米腩排　　44
Steamed Loin Ribs with
Glutinous Rice on Lotus Leaf

菊花炒三絲　　46
Stir-fried Pork and Vegetable Shreds
with Chrysanthemum

蜜汁叉燒拼法包　　49
Honey Glazed Roast Pork with Baguette

芝士百花球　　52
Cheese Stuffed Shrimp Balls

飯麵
鹹魚蝦乾菜粒炒飯　　56
Fried Rice with Salted Fish,
Dried Prawns and Vegetables

甜品
荔芋椰香西米露　　58
Taro Sago Sweet Soup
with Coconut Cream

特色秋宴

前菜
瑞士口水雞　62
Sweet and Spicy Chicken

湯羹
瑤柱冬茸羹　66
Dried Scallop and
Winter Melon Thick Soup

主菜
順德焗膏蟹鉢　69
Baked Mud Crabs in Shunde Style

百花煎釀小棠菜　72
Fried Shrimp Stuffed
Shanghainese Bok Choy

雙色炒蝦仁　76
Stir-fried Shrimps in Two Flavours

柱侯荔芋燜鴨　80
Stewed Taro and Duck in Chu Hou Sauce

飯麵
雜菌蝦籽燜伊麵　83
Braised E-Fu Noodles with
Assorted Mushrooms and Shrimp Roe

甜品
薑汁撞奶　86
Ginger Milk Pudding

春節盛宴

前菜
芥末沙律蝦　90
Wasabi Shrimp Salad

湯羹
花旗參燉竹絲雞　94
Double-steamed Black-skinned Chicken
with American Ginseng

主菜
梅子蒸花蟹　96
Steamed Blue Crabs with Pickled Plums

法式焗釀青口　100
Baked Stuffed Mussels in French Style

翡翠金銀帶子　103
Dual-flavours Scallops with Vegetables

蠔豉臘味生菜包　106
Dried Oyster and Preserved Meat
Wrapped in Lettuce

飯麵
鳳梨海鮮炒飯　109
Stir-fried Rice with Seafood and Pineapple

甜品
白玉雪影　112
Almond Sweet Soup with White Fungus
and Egg Whites

用手機掃描食譜內 QR 碼 ，
即時收看廖師傅的烹飪竅門。

豐盛賀年飯

前菜
香芹脆拌雞絲   116
Celery with Shredded Chicken

湯羹
迷你佛跳牆   118
Double-steamed Chicken Soup
with Abalones, Dried Scallops and
Fish Maw

主菜
古法蒸桂花魚   121
Steamed Mandarin Fish with
Mushroom and Jinhua Ham

西施炒蝦仁   124
Stir-fried Shrimps with Egg Whites

生財紅燒元蹄   128
Braised Pork Knuckle with
Chinese Lettuce

脆皮沙薑雞   131
Crisp Chicken with Sand Ginger

飯麵
臘味糯米飯   134
Glutinous Rice with Preserved Meat

甜品
銀耳紅蓮雪梨糖水   137
Pear Sweet Soup with
White Fungus and Lotus Seeds

清新素宴

前菜
脆皮滷豆腐   142
Deep-fried Spiced Tofu

湯羹
雪耳野菌羹   146
White Fungus and Mushroom Thick Soup

主菜
碧綠玉子環   149
Hairy Melon Rings with Egg Tofu

羅漢腐皮卷   152
Beancurd Skin Rolls with Vegetables

素瑤柱百合炒金粟   156
Stir-fried Lily Bulbs and
Corn Kernels with Vegetarian Scallop

椰香南瓜盅   159
Pumpkin Bowl with Coconut Milk

飯麵
素丁雙色炒飯   162
Stir-fried Two-colour Rice with
Diced Vegetables

甜品
花旗參桂花糕   166
Sweet Osmanthus Pudding with
American Ginseng

英文版食譜   169-216
Recipes in English

廖師傅的廚房必備四寶
The Four Essential Condiments in the Kitchen

烹調菜餚時，有些材料經常使用，但製作卻繁複，如蒜茸、薑汁酒、冬菇等，我建議讀者不妨考慮在家中預備這類材料，不用每次入廚時也花時間準備，令整個過程節省程序及時間。若你家的廚房常備以下材料，炮製本食譜菜餚宴客時，節省很多時間及工序之餘，也感輕鬆自在！

◀┈┈┈┈┤ 即時睇片

蒜茸 Finely chopped garlic

建議製作份量：300 克至 500 克
Suggested quantity: 300 g to 500 g

做法 *Method:*

1. 蒜肉去衣，與水同放入攪拌機內，打碎至幼粒，用笡箕盛起，浸在盛有清水的窩內，壓出蒜茸汁。

1. Skin the garlic. Put the garlic into water in a blender and grind them into granules. Place the garlic granules in a colander and soak them into a big bowl of water to wash. Press for the garlic juice.

做法 *Method:*

2. 隔水後，用乾布包着蒜茸，徹底擠乾水分，乾透後貯存於玻璃樽或密實膠盒內，放於雪櫃可保存 1 個月。

3. 每次使用時，用潔淨的茶匙舀出蒜茸，用於炒菜或醃味皆宜，方便快捷。

2. Drain the garlic granules and wrap them in a dry cloth. Squeeze water out thoroughly. When the garlic dries out, store in a glass jar or an airtight plastic box. Place in a refrigerator. It can be kept for 1 month.

3. When use each time, spoon out the garlic with a clean teaspoon. It is suitable for stir-frying with vegetables or marinating food. Quick and convenient!

小貼士 *Tips:*

· 攪打蒜茸時，蒜肉及水的比例是一比一。

· 若蒜茸沒有用清水洗淨，蒜茸容易發酵變成青綠色，縮短貯存的時間。

· 用布壓乾水分時，必須盡量擠壓水分，蒜茸乾透後才宜於貯存。

· The proportion of garlic to water for blending is 1:1.

· If the blended garlic is not washed with water, it will ferment easily and change to green colour. It will shorten the time for storage.

· When drying the garlic with a cloth, try squeezing out as much water as possible. Only the dried garlic is suitable for storage.

薑汁酒 Ginger juice with wine

建議製作份量：約 650 ml
Suggested quantity: about 650 ml

材料 *Ingredients:*

薑肉 300 克、冷開水 300 ml、
雙蒸酒或花雕酒 50 ml

300 g peeled ginger, 300 ml cold drinking water, 50 ml double distilled rice wine or Hua Diao wine

做法 *Method:*

1. 薑肉去皮，洗淨、切碎，放入攪拌機內，下冷開水打成薑茸，用密篩隔起。
2. 壓去薑茸水分即成薑汁，加入廚酒拌勻，貯存於玻璃瓶內，放於雪櫃可保留 1 個月。
3. 每次使用前，輕搖瓶內薑汁才倒出。

1. Peel the ginger. Rinse and chop up. Put into a blender. Add cold drinking water and blend together into ginger puree. Drain with a mesh strainer.
2. Press juice out from the ginger puree. Add cooking wine to the ginger juice. Mix well. Put into a glass jar. Store in a refrigerator. It can be kept for 1 month.
3. Gently shake the jar before pouring out the ginger juice for cooking each time.

小貼士 *Tips:*

· 薑汁酒可用於炒菜或灼菜（如芥蘭、豆苗）；滾煨材料時辟除腥異味（如草菇、海參或海味）。
· 可保留擠壓後的薑茸，做成白切雞的蘸料或炒煮配料，別浪費！

· The ginger juice with wine can be used in stir-frying or blanching vegetables (such as Chinese kale and pea sprouts). It can also help in blanching to remove the odd, fishy smell of some ingredients (like straw mushrooms, sea cucumbers or dried seafood).
· Don't waste the residues of ginger puree! It can be used as a dipping sauce for steamed chicken or a spice for cooking.

冬菇 Dried black mushrooms

建議製作份量：300 克至 500 克
Suggested quantity: 300 g to 500 g

做法 *Method:*

1. 冬菇用清水浸過面，約 3 小時至軟身，剪去蒂。
2. 用筲箕盛起，倒入生粉 2 湯匙，拌勻後洗擦冬菇表面污漬，用清水沖淨，擠乾水分，放入窩內。
3. 加入清水蓋過冬菇面，放入調味料（鹽 1 茶匙、糖 1 茶匙、生油 2 茶匙、花雕酒 1 茶匙、薑 3 片、葱 1 棵），隔水蒸約 1 小時（視乎冬菇厚薄而定，自行調整時間），待涼，放入密實盒冷藏貯存，可保存 1 至 2 星期。

1. Cover the dried black mushrooms with water. Soak for about 3 hours to soften. Cut away the stalks.
2. Put the black mushrooms in a colander. Put in 2 tbsps of caltrop starch. Mix well. Rub the black mushrooms to remove the dirt on their surface. Rinse with water. Squeeze water out. Put into a big bowl.
3. Add water to cover the black mushrooms. Put in the seasoning (1 tsp of salt, 1 tsp of sugar, 2 tsps of oil, 1 tsp of Hua Diao wine, 3 slices of ginger, 1 sprig of spring onion). Steam on boiling water for about 1 hour (adjust the steaming time according to the thickness of the mushroom). Let it cool down. Put into an airtight box and keep refrigerated. It can be kept for about 1 to 2 weeks.

小貼士 *Tips:*

· 冬菇處理後，用途甚廣，可原隻冬菇使用，或切成菇片、菇絲及菇粒等配料炒煮，隨時使用，非常方便。
· 每次使用時建議用筷子取出，別用手接觸，以免受到污染。
· 貯存時別加入太多冬菇蒸汁浸泡，或不要也可。
· The processed black mushrooms can be widely used in cooking. They can be used in whole, or sliced, shredded or diced pieces. You can use it any time. So convenient!
· It is recommended to take the black mushroom with a pair of chopsticks. Don't touch by hand to avoid contamination!
· It is not necessary to add too much steamed sauce to the black mushrooms for storage. You may simply skip the sauce.

建議製作份量：500 克
Suggested quantity: 500 g

金華火腿肉 Jinhua ham meat

做法 *Method:*

1. 將金華火腿（連皮及骨）用清水洗擦，放入滾水內略灼，去掉表面油脂，放入窩內。
2. 用熱水浸過火腿面，加入薑 1 片，隔水蒸約 1 小時，取出待涼。
3. 用刀切去火腿皮，拆掉火腿骨，火腿肉用保鮮盒或保鮮紙密封冷藏貯存，餘下的火腿蒸汁、火腿皮及骨可作煲湯之用。

1. Rub the Jinhua ham (with skin and bone on) with water. Put into boiling water. Slightly blanch to remove grease on the surface. Put into a big bowl.
2. Add hot water to cover the ham. Put in 1 slice of ginger. Steam on boiling water for about 1 hour. Let it cool down.
3. Cut away the skin with a knife. Remove the bone. Put the ham into an airtight box or wrap it tightly in cling wrap. Keep refrigerated. The steamed ham sauce, skin and bone can be reserved for cooking soup.

11

小貼士 *Tips:*

· 火腿肉冷藏貯存，保存期較長，可達 2 至 3 個月。
· 火腿肉用途很多，可切絲、切片、切條或切茸，用於各款家常小菜或宴客菜式。
· The ham kept in a refrigerator can last longer. It can be kept for 2 to 3 months.
· There is a wide usage of ham. It can be shredded, sliced, finely cut, or cut into strips for cooking a wide variety of homemade or banquet dishes.

夏天饗客

酷熱天，胃口納悶，
以冬瓜、香茅、露筍、鮮橙、綠茶當主角，
口感，賣相，涼透心！

Summer Feast

宴客：8 至 10 位享用

前菜

酥脆響鈴
Crisp Beancurd Skin Rolls

湯羹

魚肚豆腐羹
Fish Maw and Tofu Thick Soup

主菜

白玉火鴨卷
Winter Melon and Roast Duck Rolls

鮮露筍炒蝦片
Stir-fried Asparagus and Shrimp Slices

香橙脆肉脯
Deep-fried Pork in Spiced Orange Sauce

香茅鹽焗烏頭魚
Salt-baked Grey Mullet with Lemongrass

飯麵

上湯水餃麵
Dumplings and Noodles in Soup

甜品

綠茶凍糕
Cold Green Tea Pudding

酥脆響鈴
Crisp Beancurd Skin Rolls

材料:

腐皮	2 塊
豬肉碎	100 克

調味料:

清水	5 湯匙
蠔油	1 湯匙
生粉	2 茶匙
胡椒粉、麻油各適量	

購買食材 tips

- 選用薄圓形及呈半透明的腐皮（直徑約 2 呎半），一般售賣豆類製品的店舖有售，每塊約 4 至 5 元，有部份店舖出售兩片裝腐皮。買回來後用塑膠袋密封存放雪櫃，以免風乾及變壞。
- 豬肉選肥瘦均勻的梅頭肉，比瘦肉為佳。

前一天準備篇

豬肉剁碎或攪碎至幼滑，加入調味料拌勻，放雪櫃存放待明天使用。

當天宴客篇

1. 腐皮在中間剪開成兩塊半圓形，將已調味的豬肉鬆平均塗在腐皮上，捲成圓筒形，剪成約 2cm 寬的小圓卷。

   「捲時不要太緊，略鬆即可。」

2. 鑊內放入生油約 2 杯，用中油溫燒熱，放入腐皮小卷炸至金黃色，隔去油分，上碟，以唸汁蘸吃。

魚肚豆腐羹
Fish Maw and Tofu Thick Soup

材料：

乾魚肚	150 克
布包豆腐	2 件
金華火腿	10 克
雞湯	500 毫升
雞蛋白	2 個
清水	500 毫升
薑	4 片
葱	2 棵
米醋	1 湯匙

調味料：

鹽	2/3 茶匙
麻油	半茶匙
胡椒粉	適量

生粉水

清水	80 毫升
生粉	30 克

* 拌勻

購買食材 tips

- 乾魚肚分為油炸肚（圖左）
  及沙爆肚（圖右）：油炸肚
  用油炸成，色澤較淡黃，保
  存期較短，容易有油饐味，
  價格較便宜；沙爆肚是用熱
  沙炒爆而成，色澤較白，表
  面乾爽，起發質素較佳，價
  格較油炸肚略貴。

- 金華火腿可選購真空包裝的，
  重約 2 至 3 兩，價格較便宜，
  但質素比原隻火腿稍遜。

- 雞湯可購買現成罐裝或盒裝
  產品。

乾魚肚做法：

1. 乾魚肚用冷水脹發浸軟（約2小時），切成粒狀。

2. 鑊內燒熱水，加入米醋、薑片及蔥煮滾，下魚肚粒滾透（約3分鐘）去異味，盛起，用冷水漂凍，瀝乾水分，放入雪櫃貯存備用。

金華火腿做法：

1. 金華火腿用滾水略灼（約1分鐘），<u>放碗內用清水浸過面</u>，蒸約半小時，待涼後盛起。

2. 金華火腿切成2吋長方件（留作「白玉火鴨卷」之用），將剩餘的火腿肉剁成火腿茸貯存，蒸火腿水加入雞湯一併使用。

> 「火腿用水浸過面蒸，去鹹味之外，以免肉質乾硬，蒸火腿汁也可加入湯羹煲煮。」

1. 布包豆腐切粒，用滾水浸熱備用。

2. 魚肚從雪櫃取出，用滾水滾透，隔水備用。

3. 清水及雞湯放煲內煮熱，放入魚肚，豆腐隔水放進湯內，加入調味料煮滾，攪動湯水，倒入生粉水勾芡，煮滾後即可熄火，再下蛋白拌成蛋花，盛於碗內，灑上火腿茸即可。

白玉火鴨卷

Winter Melon and Roast Duck Rolls

材料：

冬瓜	1 1/2 公斤
燒鴨髀	1 隻
冬菇	5 個
甘筍	200 克
金華火腿	60 克
韭菜	300 克
菜心	1 斤（約 16 條）

芡汁：

清水	80 毫升
蠔油	1 湯匙
鹽	1/3 茶匙
雞粉	半茶匙

生粉水：

| 生粉 | 2 湯匙 |
| 清水 | 5 湯匙 |

* 拌勻

購買食材 tips

- 冬瓜最宜選外表無裂痕及光滑，瓜身重而肉厚為佳。
- 金華火腿要選淨肉為佳，可購買真空包裝。
- 燒鴨選用厚肉部位最佳，如鴨髀或鴨胸部位皆可。

前一天準備篇

- 冬瓜削皮、去核，切成5吋x3吋的長方形瓜件數塊，再直切約2mm厚的冬瓜片約20片，用保鮮紙密封，貯存雪櫃內。
- 冬菇的處理方法參考p.10，切粗條，冷藏備用。
- 金華火腿可取用「魚肚豆腐羹」的火腿件，切條（約2吋長及1cm寬），貯存備用。
- 甘筍切條，與火腿條形狀相同。

當天宴客篇

1. 冬瓜片及韭菜用滾水浸約1分鐘，盛起，用冷水漂凍，隔水備用。
2. 燒鴨髀起肉，切條（約2吋長）備用。
3. 將一片冬瓜片鋪平，放上燒鴨條、甘筍條、火腿條及冬菇條各一，捲起，用韭菜扎成卷狀，約做20件。
4. 火鴨卷放在碟上，用大火蒸約6分鐘，取出。菜心灼熟，隔水後拌碟。
5. 芡汁煮熱，加入生粉水煮滾，澆在火鴨卷即成。

鮮露筍炒蝦片
Stir-fried Asparagus and Shrimp Slices

材料：		蝦膠調味料：	
鮮露筍	800 克	鹽	2/3 茶匙
蝦仁	400 克	雞粉	1 茶匙
馬蹄肉	100 克	生粉	2 湯匙
（約 6 至 7 粒，剁碎）		麻油	1 茶匙
		胡椒粉	1/3 茶匙
		蛋白	半個

料頭：
蒜茸、薑片、甘筍花各適量

調味料：

鹽	1/3 茶匙
雞粉	半茶匙
生粉	1 茶匙
清水	3 茶匙
麻油	半茶匙

* 拌勻

購買食材 tips
- 鮮露筍別選太粗或太幼身的，以澳洲露筍最佳，內地或泰國出產的雖比澳洲的便宜，但質素較差。
- 蝦仁可選用冷藏蝦仁，在超市或凍肉店有 1kg 包裝出售，一般由越南或泰國生產。

蝦膠做法：

1. 冷藏蝦仁解凍（若用鮮蝦仁需去殼），蝦肉用生粉洗擦乾淨，沖淨。

2. 蝦仁用乾布吸乾水分，<u>用刀拍散蝦肉，再用刀背剁成茸</u>。

3. 蝦肉放入大窩內，<u>用手擦約3分鐘至有少許黏力</u>，加入蝦膠調味料順一方向攪擦（約5分鐘），至蝦肉帶黏質及彈性，用手撻至蝦肉起膠，放入馬蹄肉拌勻，冷藏備用。

> 「別用刀鋒剁蝦仁，以免切斷蝦肉纖維。」

> 「用手搓擦蝦肉，容易起膠帶黏性。」

1. 將露筍末端老根切去（約4cm），棄掉，削去硬皮，洗淨，斜刀切成厚片。

2. 鑊燒熱，下1湯匙生油，放入蝦膠用鑊鏟壓平，用慢火煎熟至兩面金黃色，隔油，用刀斜切成蝦片備用。

3. 鑊內燒滾半杯清水，加入薑汁酒1湯匙、鹽、糖、生油各1茶匙，下露筍用中火煸炒至7成熟（約1分鐘），隔水備用。

4. 鑊燒熱後，下生油1茶匙，用小火將料頭爆香，放入露筍及蝦片用中火快炒十數下，傾入調味料炒勻，最後在鑊邊潵酒，快炒數下即可上碟。

香橙脆肉脯
Deep fried Pork in Spiced Orange Sauce

材料：

豬肉眼	500 克
鮮橙	1 個
麵包糠	150 克
雞蛋	1 個

醃料：

鹽	半茶匙
雞粉	半茶匙
生粉	4 茶匙
雞蛋	1 個
濃縮橙汁	4 茶匙
清水	4 湯匙

香橙汁：

白醋	50 毫升
清水	100 毫升
濃縮橙汁	100 毫升
砂糖	60 克
鮮橙	1 個 (去皮切粒，後下)
橙酒或甜酒	2 茶匙 (後下)
吉士粉	20 克

購買食材 tips

- 肉眼可選用新鮮豬肉眼或冰鮮無骨豬扒。
- 麵包糠選用日本粗粒麵包糠較鬆脆。
- 濃縮橙汁可選用 Sunquick 牌之產品，橙味較濃郁。

肉脯做法：

肉眼切去肉筋部分，切厚片（約
0.5cm 厚），洗淨，瀝乾水分，醃料
攪勻，放入肉脯拌勻，冷藏貯存備
用。

香橙汁做法：

白醋及橙汁用慢火煮熱，加入砂糖
煮溶，吉士粉與清水拌勻勾芡，最
後加入橙酒及橙粒拌勻，待涼後冷
藏備用。

當天宴客篇

1. 已醃製的肉脯黏上麵包糠，用手壓緊以免
   脫落。
2. 鑊內燒熱生油2杯，下肉脯用中油溫炸熟
   至金黃色盛起，吸去油分，上碟，煮熱香
   橙汁，澆在肉脯上享用。

「先放一片肉脯下油
鑊，測試油溫熱度
是否足夠，若油溫不
足，麵包糠會脫落。」

香茅鹽焗烏頭魚
Salt-baked Grey Mullet with Lemongrass

材料：		佐汁料：	
烏頭魚	2條（約1斤）	魚露	3湯匙
香茅	4支	開水	3湯匙
芫茜	6棵	砂糖	1湯匙
葱	4棵	青檸	2個（榨汁）
粗鹽	2斤	白米醋	1茶匙
檸檬葉	4片	芫茜梗	3棵（切碎）
		指天椒	2隻（切粒）
		蒜茸	1茶匙

* 拌至糖溶化

購買食材 tips

市場出售的烏頭魚一般是冰鮮貨，購買時選魚身飽滿、魚鱗沒脫落、魚鰓紅潤為佳。

前一天準備篇

由於是海產類，不宜預早處理製作。

當天宴客篇

1. 焗爐調至 220℃，預熱 10 分鐘。
2. 烏頭魚毋須去鱗，洗淨，抹乾水分備用。
3. 香茅切段，在每條魚的魚腔放入香茅 2 支、芫茜 3 棵、葱 2 棵及檸檬葉 2 片。
4. 焗盤鋪上錫紙，依次鋪上粗鹽及魚，再在魚身灑上粗鹽封面，用錫紙將魚身密封，放入焗爐焗約 25 分鐘，取出。
5. 享用時，掃去魚身上的粗鹽，魚腔內香料棄去，魚腹鋪平放在碟上，蘸汁料食用。

「魚鱗可保護油脂部份以免流失。」

上湯水餃麵
Dumplings and Noodles in Soup

材料：		調味料：	
廣東生麵	4 個	鹽	半茶匙
廣東水餃皮	300 克（約 25 片）	蠔油	1 湯匙
菜心	12 兩（約 12 條）	雞粉	1 茶匙
上湯	500 毫升	生粉	4 茶匙
清水	300 毫升	清水	2 湯匙
韭黃	30 克	雞蛋白	1 個
		麻油	1 茶匙
餡料：		胡椒粉	1/3 茶匙
鮮蝦肉	20 隻		
梅頭肉	400 克		
木耳	1 朵		
蝦米	40 克		
薑	40 克		

購買食材 tips

- 廣東生麵（即鹼水麵）是麵店雲吞麵的種類，有粗麵及幼麵之分，買回後先弄散麵餅，放入袋內貯存，令生麵的鹼水味揮發掉。
- 廣東水餃皮較薄身、呈淡黃色，與略厚及白色的上海水餃皮差別很大，售賣粉麵的店舖有售。水餃皮宜密封貯存，以免被抽乾水分。
- 木耳選用白背黑木耳，體大，質感爽口。
- 上湯可選購罐裝或盒裝產品。

餡料做法：

1. 鮮蝦肉用生粉洗擦，沖水洗淨，用乾布吸乾水分，切粗粒備用。
2. 木耳浸軟，切幼絲，用滾水略灼去異味，冷水漂凍備用。
3. 蝦米用清水浸軟（約15分鐘），洗淨隔水，剁成幼粒，用乾鑊炒香備用。
4. 薑切成薑米；梅頭肉洗淨，剁碎備用。
5. 全部餡料放入碗內，下調味料攪至起膠，冷藏備用。

「水餃不宜預早包好，即包即灼，以免餡料濕潤弄破水餃皮。」

1. 將水餃餡包製成水餃（約25粒）。
2. 菜心切成菜遠，洗淨備用；韭黃切粒，洗淨備用。
3. 用滾水灼熟生麵（不要灼至過腍），盛起，沖冷水，再放入滾水內灼熱麵條，放入窩內。
4. 水餃及菜遠分別用滾水灼熟，轉放窩內。
5. 上湯及清水煮滾（如上湯鹹味不足，可加鹽1/3茶匙調味），灑上韭黃粒，上湯倒入窩內享用。

「灼水餃時，水要滾動及用隔篩輕托，以防黏着鍋底。」

材料：

綠茶粉	50 克
鮮奶	300 毫升
清水	1 杯
砂糖	70 克
魚膠粉	26 克

購買食材 tips

- 綠茶粉超市有售，一般是台灣、中國及日本出產，以日本出產之味道較濃。包裝方面有小罐裝及袋裝（有小包獨立包裝，每小包含 10 至 15 克），以袋裝較適合，剩餘後保存較佳。
- 鮮奶可選用一般盒裝鮮奶，若用忌廉的話，份量減 50 毫升，因其奶味較濃，而且比一般鮮奶香滑。

前一天準備篇

1. 魚膠粉用開水半杯浸過面，攪勻，使其吸收水分備用。
2. 綠茶粉及鮮奶拌溶，用密篩隔去未溶之粉粒。
3. 燒滾清水，下砂糖溶解，加入綠茶鮮奶拌勻。
4. 魚膠粉用微波爐以中火加熱 30 秒至溶化（亦可用熱水座溶魚膠粉），加入綠茶熱奶內攪勻，傾入甜品杯或糕盤，待涼，冷藏至凝固。

當天宴客篇

綠茶凍糕從雪櫃取出，進食時可加鮮果、淡奶或忌廉伴吃。

綠茶凍糕
Cold Green Tea Pudding

秋天宴會

夏去秋來，
選配荷葉、黃菊花瓣入饌，
荷香滿室，完美之選！

Autumn Feast

宴客：8至10位享用

前菜

麻辣蒜香鴨舌
Spicy Garlic Duck Tongues

湯羹

砂窩雲吞雞
Wontons and Chicken in Casserole

主菜

荷葉籠仔蒸糯米腩排
Steamed Loin Ribs with Glutinous Rice on Lotus Leaf

菊花炒三絲
Stir-fried Pork and Vegetable Shreds with Chrysanthemum

蜜汁叉燒拼法包
Honey Glazed Roast Pork with Baguette

芝士百花球
Cheese Stuffed Shrimp Balls

飯麵

鹹魚蝦乾菜粒炒飯
Fried Rice with Salted Fish, Dried Prawns and Vegetables

甜品

荔芋椰香西米露
Taro Sago Sweet Soup with Coconut Cream

麻辣蒜香鴨舌
Spicy Garlic Duck Tongues

材料：

急凍鴨舌	600 克
紅椒粒	30 克
白芝麻	2 茶匙

鴨舌調味料：

生蒜茸	40 克
花椒粉	1 茶匙
花椒油	1 茶匙
麻油	2 茶匙
豆瓣醬	2 茶匙
蠔油	1 湯匙
砂糖	2 湯匙

滷水調味料：

清水	2 杯
生抽	1/3 杯
蠔油	2 湯匙
鹽	1 茶匙
片糖	1 片（約 60 克）
雞粉	2 茶匙
花椒粉	1 湯匙
滷水香料	40 克
蒜茸	30 克
薑粒	20 克
乾葱粒	20 克
生油	2 湯匙
紹興花雕酒	2 湯匙（後下）

購買食材 tips

• 鴨舌是急凍貨品，一般凍肉店以獨立包裝出售。

• 花椒粉是小瓶裝的；花椒油則是大枝裝的，在大型超市及雜貨店有售。

滷水鴨舌做法：

1. 將滷水料的蒜茸、薑粒及乾葱粒用生油爆香，
   再加入其他滷水料用中火煮滾，轉小火煲約
   20分鐘，關火蓋着，待涼至滷水香料出味後，
   盛起滷水汁備用。

2. 鴨舌解凍、洗淨。煲內燒沸水，放入鴨舌煮
   滾後關火，浸焗10分鐘，盛起鴨舌，用清水
   漂凍，隔水瀝乾。

3. 鴨舌放入滷水汁內，放入雪櫃浸醃約3小時，
   盛起，冷藏備用。

「解凍時要用水
沖洗數次，使鴨
舌完全軟身，以
免有雪藏氣味。」

1. 取鴨舌調味料的蒜茸20克，用2湯匙生油用慢火炒成香蒜，
   備用。

2. 將鴨舌調味料拌勻，再加入香蒜攪拌。

3. 白芝麻放於乾鑊，用小火慢慢炒香至金黃色，待涼備用。

4. 滷水鴨舌取出，倒入鴨舌調味料拌勻上碟，最後灑上白芝麻
   及紅椒粒即可。

砂窩雲吞雞
Wontons and Chicken in Casserole

材料：

光雞	1 隻
金華火腿肉	200 克
金華火腿骨	500 克
排骨	400 克
娃娃菜	600 克
大薑片	4 片
清水	14 杯

菜肉雲吞材料：

梅頭肉	300 克
大白菜	300 克
薑粒	40 克
上海雲吞皮	24 片

雞湯調味料：

鹽	1 茶匙

雲吞調味料：

鹽	半茶匙
雞粉	半茶匙
蠔油	1 湯匙
雞蛋白	1 個
生粉	2 茶匙
清水	3 茶匙
胡椒粉及麻油	各少許

前一天準備篇

菜肉雲吞做法：

1. 梅頭肉洗淨，攪碎或剁碎。
2. 大白菜洗淨、切碎，用沸水稍灼，盛起漂凍水，瀝水後榨乾水分。
3. 大白菜碎粒、薑粒及調味料與肉碎拌勻，攪至帶膠質成餡料，用保鮮紙包好貯存雪櫃，留待翌日包製。

當天宴客篇

1. 光雞及排骨洗淨，金華火腿及排骨用熱水稍灼，去油脂及血水。
2. 大煲內加入清水 3.5 公升煲滾，先加入排骨、火腿及薑片用大火煲半小時，再用中小火煲約 1 1/2 小時成湯底，放入光雞以中火再煲約 1 1/2 小時。
3. 娃娃菜切成長條，洗淨備用。
4. 菜肉雲吞餡料取出，包成雲吞備用。
5. 將湯內之排骨取出，去骨留肉，光雞及排骨肉放入砂窩內，火腿拆成肉絲，連湯傾入砂窩用中慢火煲滾。
6. 雲吞及娃娃菜灼熟，轉放砂窩湯內，用慢火煲滾湯（約 10 分鐘）即可。

043

「菜肉雲吞毋須灼得太熟，因隨即再轉放砂窩煲煮。」

購買食材 tips

- 光雞以新鮮或冰鮮的皆可。
- 金華火腿肉可選市面淨肉真空包裝的；火腿骨以火腿腳部連皮最適合，因火腿皮帶香味。當然最好購買原隻金華火腿切件部份為上品。
- 選上海雲吞皮，比廣東雲吞皮厚身且不易弄破。
- 白菜最宜用大白菜，比較厚身且爽口。

荷葉籠仔蒸糯米腩排
Steamed Loin Ribs with Glutinous Rice on Lotus Leaf

材料：

腩排	600 克
糯米	300 克
杞子	20 克
乾荷葉	1 塊
芫茜	1 棵

料頭：

薑幼粒	10 克
蒜茸	20 克
陳皮	1 小片（切幼粒）

醃料：

豆瓣醬	1 茶匙
蠔油	2 湯匙
鹽	半茶匙
砂糖	1 茶匙
五香粉	半茶匙
廚酒（花雕酒或雙蒸酒）	1 湯匙
生粉	2 湯匙
胡椒粉	1/4 茶匙
麻油	1 茶匙
清水	4 湯匙

購買食材 tips

- 荷葉分為鮮荷葉及乾荷葉。鮮荷葉色澤鮮明，香氣較淡，夏季有售。乾荷葉外表枯殘，但香氣較濃，一般市場皆有售賣。
- 腩排肥瘦均勻，若選肉排則瘦肉較多。

前一天準備篇

- 腩排斬成 1 吋 × 2 吋長方形，洗淨，瀝乾水分。將醃料及料頭拌勻，放入腩排攪勻醃味，冷藏貯存備用。
- 乾荷葉預早購買，用膠袋封密貯存雪櫃。

當天宴客篇

1. 糯米洗淨，用清水浸過面約 3 至 4 小時，備用。
2. 杞子用清水浸軟，洗淨備用。
3. 乾荷葉用熱水稍灼，漂冷水洗淨，瀝乾水分，鋪在蒸籠或碟內備用。
4. 糯米瀝乾水分，放在窩內，放入腩排，令表面沾滿糯米，排於荷葉上，灑上杞子蒸約 25 分鐘，最後以芫茜裝飾即可。

菊花炒三絲
Stir-fried Pork and Vegetable Shreds
with Chrysanthemum

材料：		料頭：	
肉眼	200 克	蒜茸	1/3 茶匙
叉燒	200 克（切幼條）	薑絲	10 克
冬菇	10 個	青、紅甜椒絲	20 克
（做法參考 p.10、切絲）			
西芹	300 克	醃料：	
甘筍	100 克	生抽	2 茶匙
黃菊花	1 朵	糖	半茶匙
檸檬葉	1 片	生粉	2 茶匙
原味粟米片	60 克	清水	2 湯匙

滾煨調味料：		調味料：	
滾水	2 杯	鹽	半茶匙
鹽	1 茶匙	雞粉	半茶匙
砂糖	半茶匙	蠔油	1 茶匙
生油	1 茶匙	老抽	1 茶匙
		清水	3 湯匙
		生粉	2 茶匙

購買食材 tips

- 豬肉眼是淨肉部位，較其他部位瘦、沒脂肪，亦可選用肉排，去骨後切肉絲即可。
- 冬菇可選較便宜的，用來切絲，毋須選太厚身。
- 黃菊花或白菊花皆可，檸檬葉也可在花店一併購買。
- 宜選原味粟米片，味道較淡，鋪在餸菜底部，不會影響味道。

前一天準備篇

肉眼切成0.5cm厚片，鋪平再切成肉絲，洗淨，瀝乾水分。醃料拌勻，與肉絲攪勻，冷藏貯存。

「肉眼放冰箱冷藏半小時至略硬，較容易切成絲。」

當天宴客篇

1. 西芹削去硬絲，洗淨，切幼條備用。
2. 甘筍去皮，切厚片，再切幼條備用。
3. 檸檬葉洗淨，切幼絲；菊花剪去花托，花瓣用鹽水浸泡數分鐘，隔水備用。
4. 燒滾水2杯，加入滾煨調味料，用大火煮滾，加入西芹條及甘筍條滾煨約2分鐘，隔去水分備用。
5. 鑊燒熱，加生油2茶匙，放入肉絲用中火炒熟，盛起備用。
6. 鑊再燒熱，加生油2茶匙，放入料頭炒香，加入冬菇絲、西芹條及甘筍條用中火爆炒數下，加入叉燒條及肉絲拌炒十數下。
7. 調味料拌勻，逐少倒入鑊內用大火快炒均勻，最後潷酒快炒數下。
8. 碟內鋪上粟米片，排上三絲菜餚，最後灑上菊花瓣及檸檬絲即成。

蜜汁叉燒拼法包
Honey Glazed Roast Pork
with Baguette

材料：			醃料：		
梅頭肉	1 公斤		鹽	2 茶匙	
百花蜜糖	1 瓶		雞粉	2 茶匙	
生粉	120 克		砂糖	5 湯匙	
			蠔油	2 湯匙	
蒜香法包：			生抽	1 茶匙	
法包	1 條		老抽	1 茶匙	
蒜茸	30 克		海鮮醬	3 湯匙	
鹹味牛油	40 克		芝麻醬	1 湯匙	
			玫瑰露酒	2 茶匙	
叉燒蜜汁：			乾蔥茸	1 茶匙	
海鮮醬	2 湯匙		蒜茸	1 茶匙	
百花蜜糖	3 湯匙		雞蛋	1 個	
生抽	1 茶匙				
滾水	1 湯匙				
洋蔥幼粒	30 克				

購買食材 tips

- 梅頭肉可選新鮮或急凍貨品，新鮮梅頭肉較急凍的稍韌，急凍梅頭肉燒烤後，質感略鬆軟，凍肉店有售，一般包裝超過 2 公斤。
- 百花蜜糖可選燒烤用的產品。

前一天準備篇

· 醃料拌勻，用保鮮紙封密冷藏（若天氣不太熱，可放於室溫）。
· 梅頭肉切約 8 吋長 × 2.5 吋寬 × 1 吋厚長條形，漂洗乾淨，用生粉 60 克拌勻，密封冷藏至翌日醃製。
· 鑊燒熱下油 1 茶匙，下洋葱粒炒香，放入叉燒蜜汁材料煮熱，待涼後貯存雪櫃備用。

當天宴客篇

1. 梅頭肉取出，洗去生粉及血水，拌入生粉 60 克，與醃料醃 40 分鐘。

   > 「醃味不要超過 40 分鐘，否則味道過鹹。」

2. 焗爐調至高火（約 240℃）預熱 15 分鐘，底盤鋪上錫紙以免弄污。
3. 梅頭肉平放爐架上，用高火焗約 10 分鐘，反轉另一面焗 10 分鐘至表面帶微紅色及肉邊呈微燒焦，轉低火（約 120℃）烘 30 分鐘，取出，在表面塗抹百花蜜糖。
4. 焗爐調至高溫（約 250℃ 至 280℃），放入叉燒每面各烘約 5 分鐘，調至低火再烘 30 分鐘，取出，再塗上百花蜜糖，待涼。

   > 「烘焗叉燒需時共 1 1/2 小時，要注意時間的控制。」

5. 烘焗叉燒期間，將法包橫切約 1cm 厚，將蒜茸及牛油混合，塗在法包面備用。
6. 叉燒烤烘後，取出，放入法包用低火烘約 5 分鐘。
7. 叉燒蜜汁放入微波爐翻熱，備用。
8. 法包烘脆後，取出放在碟上，叉燒斜刀切片（約 1cm 厚），放在法包上，塗上叉燒蜜汁即可。

芝士百花球

Cheese Stuffed Shrimp Balls

材料：		調味料：	
蝦仁	500 克	鹽	1 茶匙
馬蹄肉	10 粒（120 克）	雞粉	半茶匙
生油	3 杯	糖	半茶匙
麵包糠	250 克	麻油	半茶匙
		胡椒粉	1/4 茶匙

芝士餡料：		生粉	2 湯匙（後下）
卡夫芝士	4 片	雞蛋白	半個（後下）
牛油	30 克		
蟹籽	20 克		
清水	50 毫升		
砂糖	2 茶匙		

前一天準備篇

芝士餡做法：

芝士片、牛油及清水放碗內蒸溶，加入砂糖及蟹籽拌勻，傾入容器內，待涼後冷藏，待凝固後切成小粒即成。

蝦膠做法：

1. 馬蹄肉洗淨，用刀拍爛後剁碎，壓去水分備用。
2. 蝦仁解凍、洗淨，用乾布或抹手紙吸乾水分，用刀拍爛蝦肉，以刀背剁成茸，放入大碗內，下調味料順一方向拌至蝦膠帶黏性，下生粉攪至起膠，再加入蛋白攪擦均勻，用手搓撻至蝦膠黏性很強即可，冷藏至翌日使用。

1. 蝦膠唧成約 1 吋直徑小圓球，釀入小粒芝士餡，輕手搓圓，放入麵包糠內沾滿表面。

2. 鑊或小鍋內下油（油分足够覆蓋百花球為準），油溫燒至中溫（約 200℃），放入百花球（若油分不足，分開數次放入百花球），炸約 1 至 2 分鐘，調低油溫，炸至蝦球浮面及表面略呈金黃色，再轉中火炸約 2 分鐘，盛起，瀝乾油分，上碟，伴沙律醬蘸吃。

購買食材 tips

- 購買冷藏蝦肉較方便及便宜。若用鮮蝦，必須去掉蝦殼，再用生粉洗擦表面污漬，洗淨，瀝乾水分，冷藏數小時後才可製作蝦膠（蝦膠冷藏後水分被抽乾，肉質收縮，更具彈性）。
- 卡夫芝士片較方便，鹹味不重，如用其他芝士則需注意味道是否太鹹。
- 使用日本麵包糠裹炸，質感較脆，色澤金黃。

秋天宴會

鹹魚蝦乾菜粒炒飯

Fried Rice with Salted Fish, Dried Prawns and Vegetables

材料：		調味料：	
鹹魚	1件（約60克）	鹽	1茶匙
蝦乾	80克	雞粉	半茶匙
菜心	10條（約400克）		
葱花	1湯匙		
薑粒	半湯匙		
白飯	600克		
雞蛋	2個		

前一天準備篇

· 鹹魚用小火煎香，切小粒，冷藏備用。
· 蝦乾用清水浸軟，洗淨，切小粒，冷藏貯存。

當天宴客篇

1. 菜心洗淨，切薄片備用。
2. 鑊燒熱下油2茶匙，放入薑粒、菜粒、鹹魚粒及蝦乾粒，用中火炒數十下至香味散發，傾入蛋液拌炒。
3. 加入白飯，轉小火炒勻蛋液，再調至中火快炒至白飯熱透，灑入調味料炒勻至溶化，最後撒入葱花，轉大火快炒數下即可上碟。

「鑊燒熱才下油，以免飯粒黏底。」

購買食材 tips

● 選用梅香鹹魚較香及惹味，方便的話可選用樽裝鹹魚醬，一般超市有售。
● 可選菜梗粗長的菜心，質感爽脆，或選用香氣濃的唐芹亦可。
● 蝦乾選表面乾爽、色澤金黃及帶香味的，海蝦乾味道鮮甜，價格比一般蝦乾略貴。

荔芋椰香西米露
Taro Sago Sweet Soup with
Coconut Cream

材料：		調味料：	
芋頭	500 克	砂糖	120 克
西米	200 克		
椰漿	1 小罐（約 160 毫升）		
清水	12 杯		

購買材料 tips

- 宜選購荔甫芋，外表乾爽、外皮完整、芋身較肥大為佳。
- 罐裝椰漿較新鮮椰漿易於處理，不容易變壞。

前一天準備篇

糖水材料不宜預早準備。

當天宴客篇

1. 清水 6 杯放入大煲煮滾，下西米攪勻，熄火加蓋焗約 10 分鐘，西米拌散，至呈透明粒狀（若西米未完全透明，開火煮滾後，熄火，焗至完全透明），盛起，<u>放入冰水內冷卻</u>，隔水備用。
2. 荔甫芋去皮，洗淨，切小粒備用。
3. 煮滾清水 6 杯，放入芋粒煲約 15 分鐘至腍，下糖調味拌溶，放入西米攪拌略煮，最後傾入椰漿即成。

「西米浸冰水後，具彈性質感。」

特色秋宴

秋意漸厚，膏蟹當造，
一席味濃的秋宴，
與客人細細回味！

Distinctive Autumn Feast

宴客：8至10位享用

前菜

瑞士口水雞
Sweet and Spicy Chicken

湯羹

瑤柱冬茸羹
Dried Scallop and Winter Melon Thick Soup

主菜

順德焗膏蟹鉢
Baked Mud Crabs in Shunde Style

百花煎釀小棠菜
Fried Shrimp Stuffed Shanghainese Bok Choy

雙色炒蝦仁
Stir-fried Shrimps in Two Flavours

柱候荔芋燜鴨
Stewed Taro and Duck in Chu Hou Sauce

飯麵

雜菌蝦籽燜伊麵
Braised E-Fu Noodles with
Assorted Mushrooms and Shrimp Roe

甜品

薑汁撞奶
Ginger Milk Pudding

瑞 士 口 水 雞
Sweet and Spicy Chicken

材料：		瑞士醬汁：	
光雞	半隻	香草瑞士汁	5湯匙
小青瓜	2條	辣椒油	2湯匙
炸花生	100克	花椒油	2湯匙
芝麻	1茶匙	花雕酒	1湯匙
		鎮江香醋	1湯匙
醃料：		豆瓣醬	3茶匙
薑	5片	花椒粉	半茶匙
蔥	2棵（切段）	麻油	1湯匙
鹽	3茶匙	薑米	1茶匙
八角	2粒	蒜茸	1湯匙
		蔥花	1湯匙
青瓜調味料：		芫茜碎	1湯匙
豆瓣醬	2茶匙	紅椒粒	1湯匙
砂糖	2茶匙		
麻油	2茶匙		
蒜茸	1茶匙		

前一天準備篇

· 光雞洗淨後，將醃料放入雞腔內擦勻，醃約1小時，用中火蒸半小時至熟，取出待涼，冷藏貯存備用。

· 瑞士醬汁混和，用玻璃瓶盛起，冷藏貯存。

1. 芝麻放在乾鑊內，用小火炒香備用。

2. 青瓜洗淨抹乾，開邊去籽，<u>用刀輕拍瓜身</u>後，切成約 4cm 長條，灑上鹽 2 茶匙拌勻醃約 5 分鐘，放入冰水洗去鹽味，瀝乾水分，下青瓜調味料拌勻，醃約 15 分鐘，上碟。

3. 從雪櫃取出熟光雞，室溫解凍（或放微波爐用小火解凍），斬件，鋪在青瓜上，瑞士汁煮熱澆在雞件上，最後灑上炸花生及芝麻即可。

> 「瓜身帶裂紋痕，令調味易於滲入。」

特色秋宴

065

購買食材 tips

• 光雞可選冰鮮雞，因此菜味道偏辣，反而不太着重雞鮮味。冰鮮雞可選購價格較高的品種，每隻約 60 至 70 元的貨品質素較佳。

• 小青瓜是指表皮帶小刺的幼身青瓜，比粗大的青瓜爽甜，適合製作涼菜。

• 花生可購買現成的炸花生，省卻工序。

瑤柱冬茸羹
Dried Scallop and Winter Melon Thick Soup

材料：

瑤柱	10 粒（約 80 克）
冬瓜	1 公斤
金華火腿肉	50 克
雞湯	3 杯
清水	1 杯
雞蛋白	2 個

調味料：

鹽	1 茶匙
胡椒粉	1/3 茶匙

生粉芡：

生粉	3 湯匙
清水	6 湯匙

購買食材 tips

- 冬瓜選外皮光滑無裂痕，或表皮帶一層白色粉末的，代表瓜身成熟。
- 瑤柱毋須原粒製作，可選購碎瑤柱烹調，價錢較原粒廉宜。
- 金華火腿肉可購買真空包裝的，價格較便宜。
- 雞湯可選用瑤柱雞湯味，罐裝或盒裝皆可。

· 冬瓜去皮洗淨,切成小件,盛起,放入薑
  2片,不用加水乾蒸半小時,待涼,<u>用刀
  背拍成冬茸</u>或用攪拌機打成茸,冷藏備
  用。

· 金華火腿肉洗淨,放在碗內,用清水蓋面
  蒸約40分鐘,用刀剁成火腿茸,冷藏備
  用,蒸火腿汁可用作煮湯羹。

· 瑤柱用清水浸過面(約1小時),蒸約
  45分鐘,待涼拆絲後冷藏,瑤柱水留為
  煮湯之用。

「用刀背拍成
茸,方便快捷。」

1. 將雞湯、蒸火腿汁、瑤柱水及清水煮沸,
   加入冬茸及瑤柱拌勻煮滾,加入調味料
   略拌。

2. 生粉芡攪勻,用湯杓邊攪動湯水,<u>邊逐
   少倒入生粉水煮至湯羹變稠</u>,滾後熄火,
   下蛋白輕輕拌勻,盛於窩內,最後灑上
   火腿茸即可。

「倒入生粉水時
別固定於同一位
置,以免生粉芡
結成粒狀。」

順德焗膏蟹鉢

Baked Mud Crabs in Shunde Style

材料：		餡料調味料：	
膏蟹	2 隻（每隻約 600 克）	鹽	1 茶匙
欖角	6 粒	雞粉	1 茶匙
梅頭肉	400 克	生粉	2 茶匙
肥肉	80 克	麻油	半茶匙
陳皮	1 片	胡椒粉	1/3 茶匙
芫茜	3 棵		
雞蛋	6 個	**調料味：**	
		鹽	半茶匙
料頭：		蠔油	1 湯匙
薑粒	20 克	清水	4 湯匙
蒜茸	1 茶匙		
葱粒	50 克		

購買食材 tips

- 宜選購膏蟹，蟹膏豐滿及甘香，最適合製作此菜。選膏蟹時可觀察蟹腹有少許蟹膏突起，代表充滿蟹膏，膏蟹在 6 月至 8 月最肥美。
- 也可選用鴨蛋代替雞蛋，鴨蛋的蛋味比雞蛋更濃。
- 由於需放焗爐烘焗，用瓦鉢盛載較適合，若瓦鉢細小，可分成兩份宴客。

前一天準備篇

· 陳皮用清水浸軟，刮去瓤。
· 陳皮、欖角及芫茜梗切幼粒，冷藏備用。
· 梅頭肉及肥肉剁碎，冷藏備用。

當天宴客篇

1. 將蟹身翻轉，用筷子戳入蟹腹至蟹沒有活動能力，
   拆解草繩，揭開蟹蓋，去內臟及清洗，蟹膏留用。

2. 將每邊蟹身斬成 3 件，蟹箝一開二（用刀拍裂），
   灑上生粉 3 茶匙。

3. 鑊燒熱下油 2 湯匙，放入料頭爆香，再下蟹件大
   火炒十數下，加入調味料炒勻，加蓋，用中火焗 2
   分鐘，熄火，將蟹件放入瓦鉢內。

   「蟹件不用全熟，
   以免烘焗時過火，
   影響肉質。」

4. 焗爐調至 200℃，預熱 10 分鐘。

5. 雞蛋拂勻，預留一個蛋液備用，將其餘蛋液、餡
   料調味料及材料（欖角、陳皮、芫茜及肉料）拌勻，
   最後加入蟹膏攪勻。

6. 蛋料傾入瓦鉢內浸過蟹件，蒸約 20 分鐘，取出（若
   有水分，可用紙吸乾）。

7. 在蟹鉢表面塗上蛋液，放入焗爐烘焗 10 分鐘至呈
   金黃色，取出，再塗上剩餘蛋液焗 5 分鐘至表面
   香脆，灑上葱花享用。

百花煎釀小棠菜
Fried Shrimp Stuffed
Shanghainese Bok Choy

材料：

蝦肉	300 克
小棠菜	10 棵
粟米粒	50 克
甘筍粒	30 克
雞蛋白	1 個

蝦膠調味料：

鹽	半茶匙
雞粉	半茶匙
生粉	3 茶匙
麻油	半茶匙
胡椒粉	1/3 茶匙
蛋白	半個

茨汁調味料：

雞湯	半杯
生粉	1 湯匙
清水	2 湯匙

購買食材 tips

- 可選用新鮮蝦或急凍蝦仁，新鮮蝦買回來需要加工處理比較繁複，急凍蝦於超市或凍肉店有 1kg 包裝出售，一般由越南或泰國出產；別買游水活蝦，難以去殼。
- 小棠菜選購菜身粗大的，比較適合烹調此菜。
- 粟米可買包裝淨粟米粒，用剩後冷藏貯存；罐裝粟米用不完難以貯藏，以致浪費。

即時睇片

特色秋宴

前一天準備篇

蝦膠做法：

1. 急凍蝦仁解凍（鮮蝦仁先去殼），蝦肉用生粉洗擦，沖洗乾淨。

2. 蝦仁用乾布吸乾水分，用刀拍散蝦肉，以刀背剁成蝦茸。

3. 蝦茸放入大窩內，用手搓擦約3分鐘至帶少許黏力，加入蝦膠調味料，順一方向攪擦（約5分鐘），直至蝦肉有黏質及彈力，撻入窩內數次至蝦肉起膠，放入粟米粒及甘筍粒拌勻，冷藏備用。

「蝦膠成功要訣是：吸乾蝦仁水分；用手搓擦足夠。」

075

當天宴客篇

1. 小棠菜削去少許菜葉，一開為二，洗淨，用滾水稍灼約2分鐘，沖冷水備用。

2. 小棠菜吸乾水分，在菜表面塗少許生粉，釀入蝦膠成小山形。

3. 鑊燒熱下生油2湯匙，將小棠菜釀蝦膠一面放入鑊中，用小火油溫煎熟。

4. 鑊內傾入雞湯，用小火略燜（約3分鐘），小棠菜排放碟上，雞湯用生粉水勾芡煮熟，澆在小棠菜上即可。

「蝦膠宜貼近小棠菜邊沿，否則蝦肉煎熟後收縮，容易脫落。」

雙色炒蝦仁
Stir-fried Shrimps in Two Flavours

材料：		菜汁蝦仁調味料：	
蝦仁	500 克	鹽	1/3 茶匙
西芹	400 克（削絲、切斜角）	雞粉	1/3 茶匙
菠菜	200 克	生粉	1 茶匙
甘筍	100 克（打汁用）	菠菜汁	1 湯匙
甘筍	100 克（去皮、切斜角）	麻油	1/3 茶匙

料頭（共兩份）：		茄汁蝦仁調味料：	
蒜茸	1 茶匙	鹽	1/3 茶匙
薑	數片	雞粉	1/3 茶匙
		砂糖	1 茶匙
焗炒料：		生粉	1 茶匙
鹽	1 茶匙	甘筍汁	1 湯匙
雞粉	1 茶匙	清水	1 湯匙
薑汁	1 茶匙	茄汁	2 湯匙
生油	1 湯匙		
開水	半杯		

購買食材 tips

● 西芹宜選表皮青綠、厚身及粗壯的，以美國西芹較爽甜。

● 鮮蝦或急凍蝦皆可。

菜汁蝦仁做法：

1. 蝦仁用生粉洗擦，沖淨，吸乾水分，分為兩份（每份 250 克），一份製作菜汁蝦仁，另一份製作茄汁蝦仁。
2. 菠菜洗淨，用滾水稍灼半分鐘，盛起，漂冷水，隔水備用。
3. 菠菜放入攪拌機內，加清水 3 湯匙打碎，用密孔篩隔渣，壓出菜汁備用。
4. 菜汁 4 湯匙、鹽 1/3 茶匙及生粉 2 茶匙混和，加入一份蝦仁拌勻醃製，冷藏備用，菜汁留用。

茄汁蝦仁做法：

1. 甘筍 100 克洗淨，切件，放入攪拌機內，加清水 3 湯匙打成茸，用密孔篩隔渣，壓出甘筍汁備用。
2. 取一份蝦仁加入甘筍汁 3 湯匙、鹽 1/3 茶匙、生粉 2 茶匙及麻油 1 茶匙拌勻，冷藏備用。

> 「菜蔬加熱後，其葉綠素才容易呈現出來。」

特色秋宴

079

1. 西芹及甘筍切後洗淨，鑊燒熱加入煏炒料，下西芹及甘筍煏炒約 2 分鐘，隔水，分成兩份備用。
2. 菜汁蝦仁及茄汁蝦仁分別用滾水灼熟，備用。
3. 鑊燒熱下油 1 茶匙，放入一份料頭用小火炒香，加入菜汁蝦仁略炒，下菜汁蝦仁調味料，轉中火快炒數下，再加入西芹及甘筍用大火炒十數下，上碟。
4. 鑊洗淨，燒熱下油 1 茶匙，放入另一份料頭用小火炒香，加入蝦仁略炒，倒入茄汁蝦仁調味料，轉中火快炒數下，再加入西芹及甘筍用大火炒十數下，上碟成雙色蝦仁。

柱候荔芋燜鴨
Stewed Taro and Duck in Chu Hou Sauce

材料：		調味料：	
米鴨	1 隻（約 2.5 至 3 斤）	柱候醬	3 湯匙
芋頭	2 1/2 斤	南乳	1 磚
酸梅	6 粒（去核）	鹽	2 茶匙
生油	2 湯匙	雞粉	1 茶匙
清水	5 杯	蠔油	3 湯匙
		老抽	3 湯匙
料頭：		冰糖碎	3 湯匙
薑片	60 克	紹興酒	2 湯匙
乾蔥粒	40 克		
蒜茸	1 湯匙		
蔥	3 棵（切段）		
八角	3 粒		
陳皮	1 片（浸軟、刮內瓤）		

購買食材 tips

- 米鴨可選用冰鮮貨，購買時留意鴨皮完整無破損，表皮呈淡黃色，鴨肉鮮紅為佳。
- 酸梅以大顆為佳，在雜貨店有售，有些以小袋包裝（每包約 3 至 4 粒）。

前一天準備篇

柱候鴨做法：

1. 米鴨去內臟，洗淨，剪去鴨尾棄掉（因帶騷味），鴨腳及鴨翼尖剪下留用（與米鴨同燜），用 2 湯匙老抽塗勻米鴨表面上色。

2. 鑊燒熱下油 1 茶匙，用慢火將米鴨煎香至金黃色備用。

3. 鑊燒熱下油 2 湯匙，下料頭用慢火爆香，加入柱候醬及南乳炒勻，下紹興酒略爆炒，加入清水 5 杯、酸梅及其餘調味料煮溶，放入米鴨煮滾後，轉慢火燜約 1 小時（期間將鴨兩面交替反轉，使鴨身受熱均勻），米鴨待涼後，連汁冷藏貯存。

當天宴客篇

1. 芋頭去皮、洗淨，切厚片（約半吋厚），用油煎成金黃色，備用。

2. 米鴨連汁放入鑊中翻熱，排入芋頭件，鴨汁剛浸過面（若鴨汁不足，可酌加開水），用慢火燜約 40 分鐘，關火焗約 10 分鐘，取出荔芋鋪在碟上，柱候鴨斬件，排於芋件上，鴨汁勾芡扒上面即成。

雜菌蝦籽燜伊麵
Braised E-Fu Noodles with Assorted Mushrooms and Shrimp Roe

材料：

伊麵餅	2 個
金菇	2 紮
鮮冬菇	60 克
秀珍菇	60 克
芽菜	150 克
蝦籽	3 茶匙
韭黃	80 克
生油	2 茶匙

調味料：

鹽	1 茶匙
蠔油	1 湯匙
雞粉	半茶匙
老抽	1 湯匙
清水	1 1/2 杯
麻油	1 茶匙

購買食材 tips

- 伊麵餅在粉麵店有售，宜買大個麵餅，麵條以粗圓、色金黃及無油饐味為佳。
- 蝦籽在海味雜貨店有散裝或小瓶裝出售。

前一天準備篇

- 一般蝦籽是生蝦籽，買回來用乾鑊慢火烘炒至香，待涼貯存備用。
- 伊麵餅用滾水灼軟身，盛起待涼，冷藏貯存備用。

> 「別過火灼至麵條太腍，稍灼軟身即撈起。」

當天宴客篇

1. 鮮冬菇及秀珍菇切粗條；金菇及韭黃切段，用水沖洗備用。
2. 鑊燒熱下生油 2 茶匙，放入蝦籽 2 茶匙及雜菌大火炒香，下調味料煮滾，加入伊麵以慢火燜約 3 分鐘。
3. 加入芽菜拌勻，燜至伊麵至腍，下韭黃拌勻，上碟，最後灑上蝦籽 1 茶匙即可。

薑汁撞奶
Ginger Milk Pudding

材料：

高鈣低脂奶	1 盒（大，約 1 公升）
生薑	400 克

調味料：

砂糖	6 茶匙

◄┈┈┈┈┤ 即時睇片

前一天準備篇

此甜品只宜即做即食，毋須提早預備。

當天宴客篇

1. 生薑去皮，洗淨，用薑磨器刨出薑茸（若用攪拌機打碎，生薑需先切片再攪碎成茸）。

2. 薑茸擠出薑汁，預備8隻小碗，每隻碗放入薑汁2湯匙，薑茸留作烹調之用。

3. 鮮奶煲至100℃煮沸（期間不斷攪動，以免黏底），下糖拌溶，關火，待鮮奶降溫約75℃至80℃，撞入小碗內（八成滿即可），靜待5分鐘至凝固，即可享用。

購買食材 tips

- 我建議選購維記高鈣低脂奶，若採用新鮮水牛奶更佳。
- 生薑以薑枝粗大、表面光滑、質地爽脆、折斷後有薑汁滲出為佳。

087

＊ 若家中沒有食物溫度計，我建議用以下方法降溫：鮮奶煮沸後，倒入另一個小煲或量杯內，再到回鮮奶煲內，來回4次後，溫度下降，即可撞入碗內，靜待5分鐘凝固即可。

春節盛宴

團圓春節，
以海鮮盛宴款客，
賓客歡欣，盡興而歸！

Spring Festival Feast

宴客：8 至 10 位享用

前菜

芥末沙律蝦
Wasabi Shrimp Salad

湯羹

花旗參燉竹絲雞
Double-steamed Black-skinned Chicken with
American Ginseng

主菜

梅子蒸花蟹
Steamed Blue Crabs with Pickled Plums

法式焗釀青口
Baked Stuffed Mussels in French Style

翡翠金銀帶子
Dual-flavours Scallops with Vegetables

蠔豉臘味生菜包
Dried Oyster and Preserved Meat
Wrapped in Lettuce

飯麵

鳳梨海鮮炒飯
Stir-fried Rice with Seafood and Pineapple

甜品

白玉雪影
（雪耳杏汁蛋白糖水）
Almond Sweet Soup with
White Fungus and Egg Whites

芥末沙律蝦
Wasabi Shrimp Salad

材料：		沙律汁：	
蝦仁	20 隻	沙律醬	200 克
青瓜	2 條	日本芥末醬	2 茶匙
雞蛋	3 個	煉奶	1 湯匙
		檸檬汁	2 茶匙

購買食材 tips

- 蝦仁選用新鮮蝦或急凍蝦仁皆可。
- 青瓜購買外表光滑、粗壯的品種較佳，因瓜肉較多及爽脆，削絲較為方便。
- 沙律醬可選用卡夫奇妙醬。

前一天準備篇

- 急凍蝦仁解凍（鮮蝦仁先去殼），蝦肉用生粉洗擦，沖洗乾淨，用鹽1茶匙醃半小時，用滾水灼熟，放入食用冰水內待涼，蝦仁開邊，用冷開水洗淨蝦腸，吸乾水分，冷藏。
- 青瓜去皮，用水洗擦，抹去水分，用削磨刨將青瓜刨絲，浸入冰水內，再下十數粒冰塊，冷藏約3小時，隔水，吸乾水分，冷藏備用。
- 雞蛋（連殼）放入冷水內煲滾至熟（約12分鐘），漂冷水待涼，去除蛋殼，冷藏備用。

當天宴客篇

1. 沙律汁拌勻，冷藏備用。
2. 將熟雞蛋的蛋白及蛋黃分開，蛋白切成幼粒；蛋黃放入密孔隔篩內，用小匙壓出幼末備用。
3. 沙律汁3湯匙、青瓜絲及蛋白粒調勻，放在碟內，鋪上蝦仁，擠上其餘沙律汁，最後灑上蛋黃末即成。

「沙律汁需冷藏1小時以上，以免使用時溶化。」

花旗參燉竹絲雞

Double-steamed Black-skinned Chicken with American Ginseng

材料：

竹絲雞	1 隻
赤肉（瘦肉）	300 克
金華火腿肉	40 克
花旗參片	30 克
清雞湯	1 杯
薑	3 片
清水	5 杯

調味料：

鹽	1 茶匙
紹興酒	2 茶匙

購買食材 tips

- 竹絲雞可採用冰鮮貨品。
- 赤肉宜選全瘦部份，以免燉湯表面出現浮油。
- 宜用金華火腿淨肉，可選真空包裝產品。
- 在藥材店選購原枝花旗參，請店員代切片，較有保證。

前一天準備篇

- 竹絲雞洗淨、斬件，用滾水灼熟，漂涼，冷藏備用。
- 赤肉及金華火腿切粒，灼熟及漂涼後，冷藏備用。
- 花旗參片用熱水 1 碗浸焗，待涼，連水冷藏。

> 「花旗參不宜直接加熱煲煮，以免其特性變質。」

當天宴客篇

1. 清水及雞湯注入燉窩內，放入雞件、金華火腿、赤肉粒及薑片，加入紹興酒拌勻。
2. 鑊內放入清水 5 杯燒滾，放入燉窩加蓋，燉約 1 1/2 小時（若鑊內滾水不足，可酌加熱水）。
3. 注入花旗參水拌勻，再燉 1 小時，取出加鹽即可食用。

梅子蒸花蟹
Steamed Blue Crabs
with Pickled Plums

材料：		調味料：	
花蟹	2隻（每隻約1斤）	酸梅	6粒
葱	2棵	冰梅醬	2湯匙
芫茜	2棵	茄汁	2湯匙
		鹽	半茶匙
		砂糖	1湯匙
		蒜茸	2茶匙
		薑粒	2茶匙
		紅椒粒	2茶匙
		生粉	2茶匙
		生油	2茶匙
		清水	4湯匙

購買食材 tips

購買花蟹時，要選蟹爪及眼部活動能力強、部份蟹爪沒脫落及異味的。
若蟹爪及蟹箝呈透明，表示肉質未成熟，水分較多，俗稱「水蟹」；
呈奶白色的則蟹肉較多，屬於佳品。

前一天準備篇

花蟹屬海產類，需要當天選購，但蒸蟹調味料可預早調配。

1. 酸梅去核、剁碎。

2. 鑊燒熱下生油2茶匙，放入蒜茸及薑粒慢火炒香，放入所有調味料拌勻，待涼，冷藏備用。

當天宴客篇

1. 花蟹略沖洗，蟹腹朝上，用尖竹筷子在蟹腹奄尖頂部插入，至蟹失去活動能力，解開草繩，揭開蟹蓋，去內臟、蟹鰓、胃及腸臟，清洗乾淨。

2. 每邊蟹身（連爪）斬成三件，蟹箝斬開兩件，用刀拍裂蟹殼，令蟹容易入味及熟透。

3. 蟹件排在碟上，將調味料塗在蟹件上，放入滾水內隔水蒸約12分鐘，取出，灑上蔥花及芫茜即成。

「可將蟹蓋反轉，與蟹件同蒸，蒸熟後再將蟹蓋覆在蟹件上，增加美觀感。」

法式焗釀青口

Baked Stuffed Mussels in French Style

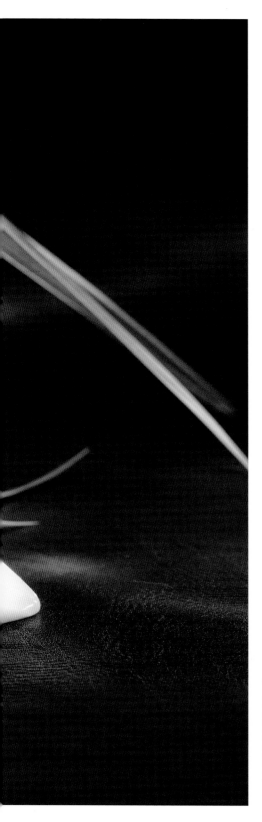

材料：

急凍青口	20 隻
豬肉碎	200 克
蘑菇	100 克
西芹	100 克
洋葱	100 克
甘筍	100 克
芝士片	4 片
雞蛋	1 個

麵粉芡汁：

牛油	60 克
麵粉	40 克
清水	4 湯匙
鮮奶	3 湯匙

調味料：

鹽	1 茶匙
雞粉	1 茶匙
糖	1 茶匙
雜香草	1 茶匙
油咖喱	1 湯匙

購買食材 tips

- 購買半殼急凍青口，一般是盒裝出售，可選購澳洲或紐西蘭出產的較有保證，每盒約 25 隻，各大超市或凍肉店有售。
- 芝士選用片裝較方便。

青口餡料做法：

1. 青口去殼起肉，擦淨青口殼留用；青口肉洗淨、切粗粒，灼熟備用。
2. 蘑菇、西芹、甘筍及洋葱切幼粒，用滾水灼熟備用。
3. 豬肉碎與鹽 1/3 茶匙拌勻，落鑊慢火炒熟備用。
4. <u>牛油落鑊慢火煮溶，加入麵粉炒香</u>，下清水及鮮奶拌勻，製成麵粉芡汁備用。
5. 鑊燒熱加入生油 1 茶匙，洋葱下鑊炒香，放入芝士片煮溶，下其他材料炒勻（雞蛋除外），加入調味料煮溶，最後下麵粉芡汁即成餡料，待涼後冷藏備用。

「別用太大火炒煮，以免焦燶，炒至金黃色即可。」

當天宴客篇

1. 焗爐預熱 10 分鐘，<u>面火約 200℃，底火 180℃</u>。
2. 青口殼用滾水灼熱，隔水，抹乾水分。
3. 餡料釀入青口殼內填滿，在表面掃上蛋液，焗約 7 分鐘，取出，再掃蛋液，焗 3 分鐘後取出，上碟享用。

「若無底面火之分，請調校溫度至 200℃。」

即時睇片

翡翠金銀帶子
Dual-flavours Scallops with Vegetables

材料：

帶子	600 克

醃料：

雞粉	1/3 茶匙
生粉	2 茶匙
麻油	1 茶匙
胡椒粉	1/4 茶匙

炸帶子（材料）：

麵包糠	200 克

蛋漿：

雞蛋	1 個
生粉	4 湯匙

* 拌勻

炒帶子（材料）：

西蘭花	600 克
鮮草菇	150 克

料頭：

蒜茸	少許
薑	數片

帶子調味料：

鹽	1/3 茶匙
雞粉	1/3 茶匙
生粉	1 茶匙
清水	1 湯匙

西蘭花調味料：

鹽	1 茶匙
雞粉	半茶匙
蠔油	1 茶匙
生粉	1 茶匙
開水	1 湯匙

購買食材 tips

- 帶子可選用急凍帶子，澳洲帶子質優但價格略高；日本急凍帶子較便宜，個子較澳洲帶子大，但品質較差。
- 宜選用日本粗粒麵包糠，炸脆後色澤金黃、口感較脆。

帶子放室溫解凍，去掉帶子邊的枕，用清水略沖洗，瀝乾水分，用醃料拌勻，分成兩份，冷藏貯存備用。

當天宴客篇

1. 西蘭花切成小朵；鮮草菇一開二，洗淨備用。

2. 取一份帶子用熱水稍灼，盛起，漂冷水，吸乾水分，<u>依次沾上蛋漿及麵包糠</u>備用。

「帶子完全沾滿蛋漿，再鋪上麵包糠。」

3. 另一份帶子用滾水稍灼，關火，利用滾水熱力浸熟（約１分鐘），隔水備用。

4. 西蘭花及草菇放入滾水內，加鹽１茶匙及油２茶匙灼熟，隔水，落鑊加西蘭花調味料用中火快炒，盛起伴碟邊，鮮草菇排在中央。

5. 鑊燒熱下油１茶匙，下料頭爆香，加入步驟３的帶子同炒，灑入炒帶子調味料，以中火快炒，讚紹酒，放在鮮草菇上。

6. <u>將沾上麵包糠的帶子放入中油溫（約180℃至200℃）</u>，炸熟至金黃色，上碟即可。

「油溫太低，麵包糠容易脫落。」

蠔豉臘味生菜包

Dried Oyster and Preserved Meat Wrapped in Lettuce

材料：		料頭：	
乾蠔豉	15 隻	薑粒	1 茶匙
臘腸	2 條	蒜茸	1 茶匙
膶腸	1 條		
梅頭肉碎	150 克	**肉碎調味料：**	
唐芹	150 克	蠔油	2 茶匙
冬菇	10 朵	生粉	1 茶匙
馬蹄肉	10 粒	清水	2 茶匙
葱	2 棵（切粒）		
粟米片	120 克	**餡料調味料：**	
西生菜	2 個	鹽	半茶匙
		雞粉	1 茶匙
燜煮調味料：		蠔油	1 湯匙
清水	2 杯	老抽	1 湯匙
鹽	1 茶匙	生粉	1 茶匙
蠔油	1 湯匙	麻油	1 茶匙
糖	1 茶匙	胡椒粉	1/3 茶匙
廚酒（花雕酒或雙蒸酒）	1 茶匙	開水	2 湯匙
薑	3 片		
生油	1 湯匙		

購買食材 tips

- 蠔豉要選外表乾爽、完整及飽滿的,以日本或南韓出產較佳。
- 臘腸及膶腸可選半肥瘦的,以外皮沒滲油及乾爽的為佳,若在表層見肥肉呈淡黃色,則表示不夠新鮮。
- 選用淡原味粟米片,超市有售。

前一天準備篇

冬菇及蠔豉做法:

1. 冬菇及蠔豉用清水浸約 3 小時至軟身,洗淨。
2. 煮滾燜煮調味料,放入冬菇慢火燜約 1 小時,下蠔豉同煮約 10 分鐘,盛起,待涼後切幼粒,冷藏備用。

臘腸及膶腸做法:

臘味用滾水稍灼,去掉表面油污,蒸約 8 分鐘,取出,待涼後切幼粒,冷藏備用。

當天宴客篇

1. 西生菜撕開菜葉,浸水洗淨,剪成盞形菜片,抹乾水分。
2. 唐芹洗淨,切幼粒;馬蹄肉剁碎,擠乾水分備用。
3. 肉碎與調味料拌勻,鑊燒熱下油 2 茶匙,下肉碎用中火炒熟,盛起備用。
4. 鑊燒熱下油 2 茶匙,放入料頭及臘味粒,用慢火炒香至滲出油分,下蠔豉粒及肉碎炒香,轉中火,加入唐芹、冬菇及馬蹄肉快炒至熱。
5. 餡料調味料拌勻,慢慢加入鑊內炒勻,最後潷紹酒,上碟。
6. 粟米片鋪在碟內,排上餡料,灑上葱花,享用時以生菜片包吃,或塗少許海鮮醬佐食。

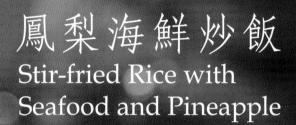

鳳梨海鮮炒飯
Stir-fried Rice with Seafood and Pineapple

材料：		醃料：	
鮮菠蘿（大）	1 個	鹽	1/3 茶匙
蝦仁	10 粒	生粉	2 茶匙
鮮魷魚	1 隻	麻油	1 茶匙
急凍帶子	10 粒	胡椒粉	1/4 茶匙
蟹柳	5 條		
甘筍	40 克	調味料：	
菜心	8 條	鹽	1 茶匙
葱	1 棵	雞粉	1 茶匙
雞蛋	2 個	生抽	2 茶匙
生油	1 湯匙		
白飯	4 碗		

購買食材 tips

• 菠蘿要選外表金黃、果身鱗目較大、有香氣散發的為佳。若見菠蘿頂葉脫落、氣味帶酒酵味，代表果身成熟，甚至有腐爛的可能性。

• 蝦仁及帶子可選急凍貨品；鮮魷魚宜選眼部突出、肉色亮澤、呈微透明、表皮不脫落者為鮮品。

· 鮮菠蘿削去綠葉，直切成兩半，用小刀或刮果肉器取出果肉（小心別弄破菠蘿殼），果肉及果殼分別用冷鹽水浸泡約 10 分鐘，隔水，抹乾水分，用保鮮紙密封冷藏，以免果肉氧化變色。

· 海鮮料解凍及切粒；鮮魷魚破開魚腹，去內臟及外皮，洗淨、切粒，海鮮料抹乾水分，下醃料拌勻，冷藏備用。

1. 蟹柳及菠蘿果肉切粒；甘筍、菜心切薄片，洗淨；葱切成葱花備用。

2. 菠蘿果殼由雪櫃取出，用溫水浸暖，抹乾水分備用。

3. 海鮮料放入熱水內灼熟。

4. 鑊燒熱下生油 1 湯匙，放入菜片及甘筍片炒香，下蛋液略炒，倒入白飯炒勻。

5. 加入海鮮料及菠蘿肉，用中火快炒至熱透，灑入調味料用大火快炒十數下至均勻即可，最後灑上葱花略炒，盛於菠蘿殼享用。

「海鮮料不宜過熟，以免炒飯時收縮過火。」

白玉雪影
Almond Sweet Soup with White Fungus and Egg Whites

材料：		調味料：	
南杏	300 克	冰糖	250 克
北杏	40 克		
乾雪耳	40 克		
清水	6 杯		
雞蛋白	8 個		
鮮奶	半杯		
薑	3 片		

購買食材 tips

- 南杏較北杏大及飽滿，杏仁香味較北杏淡，但北杏味較苦，因此配搭少量北杏以增加杏仁香味。
- 雪耳宜選漳州雪耳較佳（見圖前方），以大朵、色澤微黃、雪耳球卷緊密為佳；但現時市面有些人工培植的雪耳比漳州雪耳更大（見圖後方），色澤淡黃帶白，雪耳球卷疏鬆，浸發後呈軟身，煲煮時容易稀爛。

前一天準備篇

- 杏仁洗淨，用清水 2 杯浸約 2 小時，連水放入攪拌機打碎，用密篩隔去杏仁渣，壓出杏汁留用，冷藏貯存。
- 雪耳用清水浸軟約 2 小時，剪去黃色硬蒂，洗淨，用滾水加薑汁滾透 10 分鐘，盛起，漂冷水待涼，剪成小塊，貯存雪櫃備用。

「用薑汁可去除雪耳之異味。」

當天宴客篇

1. 煮滾清水 4 杯，加入雪耳及薑片轉小火煮約 40 分鐘，注入杏汁煮約 10 分鐘，加入冰糖調味至溶，關火。
2. 蛋白拂匀，慢慢加入糖水內攪匀，最後拌入鮮奶即可宴客。

豐盛賀年飯

新春賀年，互相恭賀，
一桌盤滿鉢滿的意頭菜，
吃得笑逐顏開！

Abundant New Year Dinner

宴客：8 至 10 位享用

前菜

香芹脆拌雞絲
Celery with Shredded Chicken

湯羹

迷你佛跳牆
Double-steamed Chicken Soup with
Abalones, Dried Scallops and Fish Maw

主菜

古法蒸桂花魚
Steamed Mandarin Fish with Mushroom and Jinhua Ham

西施炒蝦仁
Stir-fried Shrimps with Egg Whites

生財紅燒元蹄
Braised Pork Knuckle with Chinese Lettuce

脆皮沙薑雞
Crisp Chicken with Sand Ginger

飯麵

臘味糯米飯
Glutinous Rice with Preserved Meat

甜品

銀耳紅蓮雪梨糖水
Pear Sweet Soup with White Fungus and Lotus Seeds

香芹脆拌雞絲
Celery with Shredded Chicken

材料：

雞扒	3 件
西芹	300 克
甘筍	100 克
芫茜	4 棵
炸麵（油條）	1 條
紅尖椒	1 隻
唐生菜	3 片
炒香白芝麻	1 茶匙

涼拌調味汁：

豆瓣醬	1 湯匙
鹽	半茶匙
砂糖	1 湯匙
蒜茸	1 茶匙
麻油	1 湯匙
鎮江醋	2 茶匙
開水	1 湯匙

醃料：

鹽	半茶匙
雞粉	1/3 茶匙
生粉	3 茶匙
清水	3 湯匙

購買食材 tips

也可用急凍雞扒，去骨雞髀肉較方便，而且肉質厚。

前一天準備篇

· 雞扒去皮及脂肪，洗淨，下醃料拌勻待約 1 小時。
雞扒用滾水浸過面，<u>慢火浸約 20 分鐘至熟</u>，用凍開水浸涼，抹去水分，撕成雞絲，冷藏備用。

· 西芹及甘筍切幼條，洗淨，用滾水稍灼 1 分鐘，漂冰水至涼，隔水冷藏備用。

· 炸麵分成兩條，切薄片備用。

> 「以慢火浸煮雞肉，以免過火令肉質粗糙。」

當天宴客篇

1. 芫茜洗淨，切小段；紅尖椒切絲備用。

2. 唐生菜洗淨，鋪在碟上。

3. 炸麵薄片用中油溫炸脆，盛起，吸去油分備用。

4. 涼拌調味汁混和放入窩內，下西芹、甘筍及雞絲拌勻，加入脆炸麵片、芫茜及紅椒絲拌勻，上碟，最後灑上白芝麻即成。

迷你佛跳牆
Double-steamed Chicken Soup with Abalones, Dried Scallops and Fish Maw

材料:			調味料:	
原粒瑤柱	10 粒		鹽	1 茶匙
花膠筒	3 個		花雕酒	1 茶匙
急凍海參	1 條			
乾冬菇	10 朵			
罐裝湯鮑仔	1 罐			
金華火腿	80 克			
豬腱	300 克			
竹絲雞	1 隻			
大白菜	400 克			
薑	2 片			
清水	5 杯			
雞湯	2 杯			

購買食材 tips

- 可選薄身花膠筒（每斤約 20 至 24 隻），價格較便宜，而且浸發時容易處理，選外表金黃、筒圓及裂痕少為佳。
- 急凍海參大多已浸發，易於處理之餘，價格相宜。挑選時以參體飽滿、粗壯及肉厚者為佳。購買已浸發海參，以每條計約 1 斤至 1 1/2 斤較適合。

購買食材 tips：

- 罐裝湯鮑可選用湯鮑仔，每罐約 10 至 12 隻為適合，若太小則欠嚼感。
- 竹絲雞及豬腱的脂肪少，燉湯時不會出現太多油脂。
- 瑤柱可選 S 級（細元貝，每斤約 100 至 120 粒），以色澤金黃、外表完整及帶香氣為佳。

前一天準備篇

- 瑤柱用清水浸過面至漲發，連水冷藏備用。
- 花膠筒用清水浸軟（約 3 小時），隔水。沸水注入大煲內，浸焗花膠至水冷，漂冷水清洗花膠表面雜質，用薑汁滾煨去其異味，漂冷水，隔去水分，切件，冷藏備用。
- 急凍海參解凍，用薑汁酒煨透，漂冷水，切小件，冷藏備用。
- 冬菇用清水浸約 3 小時，去蒂，用生粉擦勻，洗淨冷藏備用。

> 「一般乾貨海味漲發時，宜先用冷水浸透才加熱處理。」

當天宴客篇

1. 大白菜切去大葉部份，一開四，切成菜膽，洗淨，飛水備用。
2. 竹絲雞洗淨、斬件；豬腱及金華火腿洗淨，切粒，飛水及洗淨。
3. 燉窩內鋪入雞件、火腿粒、豬腱及薑片，菜膽圍邊，冬菇、海參、花膠及湯鮑仔鋪面。
4. 雞湯、浸瑤柱水及清水 5 杯煮滾，加入紹酒 1 茶匙拌勻，注入燉窩內，排上原粒瑤柱，加蓋，用慢火燉約 4 小時，最後灑鹽調味。

* 此湯最宜用大燉窩隔水燉製，若家中沒有燉窩，可用湯鍋以慢火煲成，但湯水不及燉湯清澈。

古法蒸桂花魚
Steamed Mandarin Fish with Mushroom and Jinhua Ham

材料：		調味汁：	
桂花魚	1 條（約 1 斤）	生抽	4 湯匙
赤肉（瘦肉）	50 克	老抽	2 茶匙
冬菇	3 朵	清水	4 湯匙
金華火腿肉	20 克	雞粉	半茶匙
陳皮	1 片	砂糖	3 茶匙
薑絲	10 克	麻油	半茶匙
葱	1 棵	胡椒粉	1/4 茶匙
芫茜	2 棵	芫茜梗	2 棵

醃料：	
鹽	1/4 茶匙
蠔油	1 茶匙
生粉	1 茶匙
麻油	1 茶匙
清水	2 茶匙
胡椒粉	1/3 茶匙

購買食材 tips
- 桂花魚宜選鮮活游水的，不建議購買冰鮮貨。
- 陳皮毋須選太貴的，普通品種即可。
- 生抽可選用海鮮生抽，味道不會太鹹。

前一天準備篇

· 赤肉切成肉絲，加入醃料拌勻，冷藏備用。
· 金華火腿肉放入滾水內，以慢火煮25分鐘，隔水待涼，切絲。
· 冬菇浸軟，去蒂，用生粉洗淨，煮約25分鐘，待涼，冬菇切絲，冷藏。
· 陳皮用水浸軟，刮去內瓢，切幼絲備用。

當天宴客篇

1. 桂花魚去內臟、洗淨，抹乾水分，在魚背厚肉部分直剝一刀備用
2. 葱洗淨，切絲；芫茜洗淨，取葉留用，芫茜梗用作調味汁之用。
3. 調味汁煮熱至砂糖溶解，備用。
4. 將肉絲、火腿絲、冬菇絲、陳皮絲及薑絲拌勻，平均排放在魚身上。
5. 鑊中放入滾水4杯，放入魚加蓋以大火蒸約10至12分鐘，取出，倒出多餘水分，灑上葱絲及芫茜，澆上滾油1湯匙及調味汁即成。

123

「魚背剝開令厚魚肉容易熟透，記緊切口不要太深，否則影響外觀。」

西施炒蝦仁
Stir-fried Shrimps with Egg Whites

材料：

西蘭花	600 克
蝦仁	400 克
雞蛋白	10 個
鮮奶	150 毫升
金華火腿肉	20 克

蛋白調味料：

鹽	半茶匙
雞粉	1/3 茶匙
胡椒粉	少許
生粉	3 茶匙
清水	4 茶匙

醃料：

鹽	半茶匙
雞粉	半茶匙
生粉	3 茶匙
麻油	1 茶匙
胡椒粉	1/4 茶匙

西蘭花調味料：

鹽	1 茶匙
雞粉	半茶匙
生粉	1 茶匙
清水	2 茶匙

購買食材 tips

- 選新鮮蝦或急凍蝦仁皆可，新鮮蝦買回來需要加工處理，比較繁複；急凍蝦仁有 1kg 包裝出售。
- 金華火腿肉的用量較少，可選用真空包裝的。
- 可採用忌廉代替鮮奶，奶味濃，可視乎個人口味而定。

前一天準備篇

· 蝦仁解凍，用生粉洗擦，沖淨，吸乾水分，下醃料拌勻冷藏。
· 金華火腿放入滾水內，以慢火煮約 25 分鐘，盛起待涼，剁成茸備用。

當天宴客篇

1. 西蘭花切成小棵，洗淨備用。
2. 將蛋白調味料中的生粉及清水拌溶，混合其他調味料，傾入鮮奶內拌勻，下蛋白輕攪備用。
3. 鑊內燒熱水 2 杯，加入西蘭花及鹽 1 茶匙灼至 8 成脸，盛起，加調味料炒香，鋪在碟上。
4. 蝦仁飛水，用中油溫泡油，盛起。
5. 將鮮奶蛋白混合物拌勻，倒入鑊內，用剩油以慢火將蛋白炒至 8 成熟，放入蝦仁同炒至全熟，上碟，灑上金華火腿茸即可。

「生粉不能直接加入蛋白內，因生粉在蛋白中難以溶解，容易出現粉粒狀。」

材料：

元蹄	1 隻（約 2 至 2 1/2 斤）
唐生菜	600 克

調味料：

鹽	2 茶匙
雞粉	1 茶匙
蠔油	3 湯匙
冰糖	20 克
老抽	1 湯匙
花雕酒	1 湯匙
八角	4 粒
陳皮	1 片
薑	3 片
葱	2 棵
蒜子	4 粒
清水	4 杯

生財紅燒元蹄

Braised Pork Knuckle
with Chinese Lettuce

元蹄做法：

1. 元蹄洗淨，用滾水煮約 15 分鐘，沖洗，抹去水分，用 2 湯匙老抽塗勻表皮上色，鑊內下生油 3 湯匙，<u>放入元蹄用中火煎香表皮至金黃色。</u>

2. 鑊內用中火爆香薑、葱及蒜子，加入其餘調味料及清水煮滾，傾入大窩內，<u>排入元蹄，隔水蒸約 2 1/2 小時至腍</u>，待涼，冷藏貯存。

> 「煎元蹄時，脂肪及水分受熱容易爆開，小心濺傷。」

> 「元蹄蒸或燜皆可，以燜煮較入味，但要注意水的份量。」

130

1. 元蹄隔水蒸熱（約 20 分鐘），放在窩碟內，預留燜汁 1 杯作勾芡用。

2. 唐生菜切去菜頭，洗淨，灼熟，放在元蹄旁。

3. 煮滾燜汁，用生粉 3 茶匙與開水拌勻勾芡，扒在元蹄上，大宴親朋。

購買食材 tips

元蹄是豬的膝部，又稱為肘子，肉檔有售，一般可預定，購買時說明是去骨肘子或元蹄，店員也可代燒掉幼毛，回家後清洗即可。

脆皮沙薑雞
Crisp Chicken with Sand Ginger

材料：		上皮水：	
光雞	1 隻（約 2 1/2 斤）	白醋	4 湯匙
葱	3 棵	麥芽糖	1 湯匙
蒜子	4 粒		
鮮沙薑	50 克	**蘸汁：**	
		沙薑粉	2 湯匙
醃料：		麻油	1 湯匙
鹽	3 湯匙	熱生油	2 湯匙
雞粉	1 湯匙	鹽	半茶匙
砂糖	1 湯匙	雞粉	1/3 茶匙
沙薑粉	3 湯匙	* 拌勻	

購買食材 tips

- 光雞可選用冰鮮雞，以嘉美雞較有名及具保證，肉質及味道與鮮雞相近。購買冰鮮雞時注意其包裝要完整密封，並貯存於 0℃ 至 4℃ 雪櫃。
- 鮮沙薑氣味較沙薑粉香濃，外形細小及連皮，兩種材料皆在東南亞食品雜貨店有售。有雜貨店出售的沙薑粉，是小包裝已調味的鹽焗雞粉，味道較鹹，若用此則毋須另加調味料了。

沙薑雞做法：

1. 上皮水混合，用熱水座溶備用。

2. 光雞解凍，洗淨，抹去水分。

3. 葱切段；鮮沙薑洗淨，切粗粒，與蒜子同拍碎，將以上材料混合放入雞腔內。

4. 醃料拌勻，在雞腔內及外皮塗勻，醃約 1 小時。

5. 燒滾水，放入已醃味的光雞，隔水蒸約 25 分鐘，取出雞腔內配料棄去，待涼約半小時。

6. 用吸油紙去掉雞皮表面油脂及水分，外皮塗抹上皮水，掛起光雞晾乾至外皮乾爽（約 4 至 5 小時），將光雞用保鮮紙包好，冷藏貯存。

「將雞掛起晾乾是最有效的方法，若晾掛不便，放架上吹乾亦可。」

1. 將雞斬成兩邊，在室溫解凍約 1 小時。

2. 鑊內下生油 5 杯，用中火燒熱，轉中小火，下半份光雞炸至表面微金黃色（約 5 分鐘），轉大火燒熱油溫再炸 5 分鐘至表面金黃色，瀝去油分，放入另一半雞以相同方式炸香，斬件上碟，以蘸汁伴吃。

豐盛賀年飯

133

臘味糯米飯
Glutinous Rice with Preserved Meat

材料：

糯米	400 克
臘腸	2 條
膶腸	1 條
臘肉	半條
冬菇	8 朵
蝦米	150 克
芫茜	1 棵（切粒）
葱	1 棵（切粒）
雞蛋	1 個

調味料：

生抽	2 湯匙
老抽	2 湯匙
蠔油	1 湯匙
雞粉	1 茶匙
開水	3 湯匙
麻油、胡椒粉	各少許

購買食材 tips

● 選購臘味時，以瘦肉多肥肉少、乾爽的為佳。
● 蝦米宜選外表金黃、乾爽、蝦身完整及帶鮮香味的。

糯米做法：

1. 糯米洗淨，用清水浸過面，浸約 3 小時，隔去水分。

2. 在蒸籠或不鏽鋼笪箕鋪上乾淨紗布，鋪平糯米，隔水乾蒸約 30 分鐘成糯米飯，取出待涼，用保鮮紙包密，放入雪櫃貯存。

臘味、蝦米、冬菇做法：

1. 臘味用熱水稍灼，隔水蒸熟（約 10 分鐘），切粒，冷藏備用。

2. 蝦米用清水浸軟，切幼粒備用。

3. 冬菇浸軟，去蒂，用生粉洗擦，沖水洗淨，放入清水內，加鹽 1 茶匙調味，滾煮約 20 分鐘，切粒，冷藏備用。

1. 雞蛋拂勻，易潔鑊用小火燒熱，傾入蛋液均勻鑊面，用小火煎成蛋皮，切成蛋絲。

2. 大窩內注入溫水（約 45℃），放入糯米飯，用手弄散飯粒及洗去部分黏質，用笪箕瀝乾水分。

3. 鑊燒熱，加入生油 1 茶匙，放入臘味料用小火炒香至油分溢出，下蝦米及冬菇炒勻，加入糯米飯用中火快炒至熱，拌勻調味料逐少加入炒勻。

4. 最後轉大火快炒十數下，上碟，灑上葱花及芫茜，再鋪上蛋絲即可。

「切勿將糯米飯浸水太久，影響飯粒質感。」

銀耳紅蓮雪梨糖水
Pear Sweet Soup with White Fungus and Lotus Seeds

購買食材 tips

- 雪梨外表黃中帶綠、有果蒂、質地結實較佳。若果身呈褐色，表示過熟或開始腐爛，不宜選購。
- 選購去核紅棗，較方便處理。
- 購買原粒白湘蓮（去衣湖南蓮子），易於處理之餘，味道也較香。
- 漳州雪耳較佳，外表比人工培植的較細、帶黃，質地爽脆，加熱時不易稀爛。

材料：

雪梨	5 個
雪耳	3 朵
紅棗	20 粒
蓮子	100 克
薑	4 片
清水	8 杯

調味料：

冰糖	250 克

前一天準備篇

- 蓮子用清水浸軟（約 2 小時），以竹籤挑去蓮子芯，以免味道苦澀，清洗後冷藏貯存。
- 雪耳用水浸過面至發漲（約 3 小時），剪去硬蒂，洗淨，放入滾水內，加薑汁酒 1 湯匙滾透（約 5 分鐘）去其異味，盛起，剪碎，待涼冷藏。

當天宴客篇

1. 紅棗、薑片洗淨備用。
2. 煮滾清水 8 杯，加入雪耳、紅棗、蓮子及薑片煲滾，轉小火至微滾，煲約半小時。
3. 雪梨去皮、去芯，果肉切成粗條，洗淨，放入煲內煮 40 分鐘，下冰糖調味至溶，即可享用。

> 「冰糖不宜太早放入，當水分蒸發後，會令糖水的甜度過濃。」

清新素宴

賣相精緻的素菜，
加上創意的手藝，
引得人…胃口大開！

Fresh Vegetarian Feast

宴客：8 至 10 位享用

前菜

脆皮滷豆腐
Deep-fried Spiced Tofu

湯羹

雪耳野菌羹
White Fungus and Mushroom Thick Soup

主菜

碧綠玉子環
Hairy Melon Rings with Egg Tofu

羅漢腐皮卷
Beancurd Skin Rolls with Vegetables

素瑤柱百合炒金粟
Stir-fried Lily Bulbs and
Corn Kernels with Vegetarian Scallop

椰香南瓜盅
Pumpkin Bowl with Coconut Milk

飯麵

素丁雙色炒飯
Stir-fried Two-colour Rice with Diced Vegetables

甜品

花旗參桂花糕
Sweet Osmanthus Pudding with American Ginseng

脆皮滷豆腐
Deep-fried Spiced Tofu

材料：		佐料汁：	
布包豆腐	3 件	白醋	3 湯匙
榨菜	1 個（或 1 包）	溫水	3 湯匙
		砂糖	1 湯匙
滷水料：		蒜子	2 粒
滷水香料	40 克		
鹽	1 湯匙		
清水	2 杯		

購買食材 tips

- 滷水香料在一般藥材店有售，購買 $10 約有 80 克左右（約 2 兩）。
- 豆腐最宜購買布包豆腐，質感比板豆腐嫩滑，而且外形美觀。
- 購買原個榨菜，若節省工序可在超市或雜貨店購買包裝榨菜絲（每包約 80 克）。

滷豆腐做法：

1. 滷水香料用清水 2 杯蒸約半小時，使香料味道散發。

2. 放入鹽 1 湯匙及布包豆腐，用中火蒸約 1 小時，至豆腐表面略為粗糙，取出，隔水待涼。

3. 滷豆腐涼後，用手輕壓水分，用廚房紙或布抹乾，放入密實盒內，冷藏貯存，留待明天炸製。

榨菜片做法：

1. 榨菜洗淨表面醃料，切薄片，用冷水浸約 10 分鐘，瀝乾水分。

2. 榨菜片加入砂糖 2 茶匙、麻油 1 茶匙拌勻，放入雪櫃貯存備用。

* 如用包裝榨菜絲則毋須預先處理，直接烹調。

「原個榨菜味鹹，用糖及麻油拌勻，可減掉鹹味。」

145

1. 先處理佐料汁：蒜子剁成茸，與其餘材料拌勻。

2. 鑊內放入生油 3 杯，用大火燒熱後轉中火，放入滷豆腐炸約 3 分鐘，再轉大火炸約半分鐘，至滷豆腐表面金黃色即盛起，隔去油分。

3. 榨菜片鋪在碟內，每件炸豆腐切約 6 片，放在榨菜片上，蘸佐料汁伴吃。

雪耳野菌羹
White Fungus and Mushroom Thick Soup

材料：		齋湯材料：	
雪耳	2 朵	甘筍	500 克
冬菇	10 朵	乾冬菇	30 朵
秀珍菇	120 克	大豆芽菜	300 克
金菇	2 紥	馬蹄肉	20 粒
甘筍	100 克	鮮草菇	300 克
粉絲	2 小包	薑	4 大片
齋湯	5 杯	清水	12 杯

調味料：		生粉水：	
鹽	1 1/2 茶匙	生粉	2 湯匙
素雞粉	1 茶匙	清水	4 湯匙
老抽	1 茶匙	*拌勻	

購買食材 tips

- 宜選色澤淡黃的雪耳，以漳州雪耳為佳，不宜選顏色較白的產品。
- 金菇可選菇身較粗、菇傘軟大的。
- 馬蹄最宜購買原粒、未去皮的；真空包裝已去皮的馬蹄肉，質素較差。

前一天準備篇

齋湯做法：

1. 冬菇用清水浸軟（約 1 小時），剪去菇蒂，用生粉洗擦冬菇及菇蒂，洗淨備用。

2. 馬蹄去皮，洗淨；甘筍去皮，切片備用。

3. 鮮草菇放入滾水內，加入薑汁 1 湯匙，灼約 5 分鐘，盛起，用水沖洗備用。

4. 燒滾清水 12 杯，放入所有材料轉慢火煲約 1 1/2 小時，將冬菇、草菇及馬蹄取出（作為素宴菜餚材料），齋湯待涼後冷藏備用。

「加薑汁可去掉異味。」

＊薑汁做法：

將薑肉 300 克及清水 300 ml 攪碎，取出擠乾水分成薑汁，放於密實瓶冷藏，可保存約一個月。

當天宴客篇

1. 雪耳用清水浸軟至發漲（約 1 小時），剪去黃色硬蒂，洗淨後用滾水及薑汁 1 湯匙，煲約 30 分鐘，盛起後用水沖淨，剪碎備用。

2. 冬菇（取齋湯材料）、秀珍菇及甘筍切成幼絲備用。

3. 粉絲用清水浸軟，剪成小段，隔水備用。

4. 金菇切去尾段，鬆散後洗淨備用。

5. 煮滾齋湯，放入所有材料，轉中火煮約 5 分鐘，加入調味料，用湯杓快速攪動，慢慢下生粉水拌勻，煮滾即可。

碧綠玉子環
Hairy Melon Rings
with Egg Tofu

材料：		瓜環調味料：	
節瓜	4 個	鹽	I 茶匙
玉子豆腐	4 條	糖	I 茶匙
西蘭花	600 克	油	I 茶匙
		廚酒（米酒或紹興酒）	I 茶匙

西蘭花調味料：		芡汁調味料：	
鹽	I 茶匙	鹽	I/3 茶匙
糖	I 茶匙	素雞粉	半茶匙
油	I 茶匙	素蠔油	I 茶匙
廚酒	I 茶匙	老抽	I/3 茶匙

購買食材 tips
- 節瓜最宜挑選直身、瓜身不要太粗的，直徑約 2 吋至 2.5 吋最佳。
- 西蘭花以菜花緊密、色澤青綠為佳。

節瓜環做法：

1. 節瓜去皮洗淨，直切成約 2cm 圓厚件，呈棋子形狀，再用圓模或小刀在瓜環中央轉一圈，挖成空洞（約 2.5cm 直徑）。

2. 鑊內放入清水 3 杯煮滾，下瓜環及調味料用中火灼約 5 分鐘，至 6 成軟脸，<u>隔水後浸冰水至涼</u>，盛起備用。

3. 玉子豆腐直切成瓜環厚度，套入玉環內，放入圓碟上，貯存於雪櫃備用。

「浸冰水的作用是避免瓜環因受熱而轉黃色。」

1. 西蘭花切成細朵，洗淨備用。

2. 齋湯半杯煮熱（齋湯做法參考 p.148），傾入玉子瓜環的碟內，加蓋蒸約 10 分鐘。

3. 小煲內下熱水 2 杯，加入西蘭花調味料，下西蘭花灼約 4 分鐘，盛起備用。

4. 取出玉子瓜環，傾出齋湯放鑊內，拌入芡汁調味料煮成芡汁，澆在玉子瓜環上，以西蘭花伴碟即可。

羅漢腐皮卷
Beancurd Skin Rolls with Vegetables

材料：		調味料：	
腐皮	2 大塊	清水	1/4 杯
冬菇	10 朵	鹽	1 茶匙
蘑菇	10 粒	素雞粉	半茶匙
草菇	150 克	素蠔油	1 湯匙
馬蹄肉	10 粒	麻油	1 茶匙
西芹	120 克		
甘筍	100 克	**芡汁：**	
粉絲	1 小紮	生粉	3 茶匙
木耳	1 朵	清水	3 湯匙
		* 拌勻	

*冬菇、草菇及馬蹄肉選用齋湯材料，參考 p.148。

購買食材 tips

- 腐皮宜選直徑約 2 呎半、薄圓形及呈半透明的，一般在售賣豆類製品的店舖有售，每塊約 4 至 5 元，深水埗樹記腐竹店的品質較佳。
- 腐皮買回來後，要用塑膠袋密封存放雪櫃，以免風乾及變壞。
- 木耳選用白背木耳較爽脆。

前一天準備篇

齋餡料做法：

1. 粉絲用清水浸軟，剪成小段，隔水備用。
2. 冬菇、西芹（去皮）、甘筍及木耳切幼條備用。
3. 蘑菇、草菇及馬蹄肉切薄片備用。
4. 將所有齋料用熱水滾透約 3 分鐘，隔水盛起。
5. 鑊燒熱下油 2 茶匙，放入齋料略炒十數下，放入調味料炒勻，慢慢加入芡汁拌勻，待涼，冷藏貯存備用。

> 「餡料要宴客當天包裹，為免腐皮濕潤，即包即煎。」

當天宴客篇

1. 整塊腐皮剪成 10 小塊（約 4.5 吋 x 6 吋的長方形），放入袋內以免風乾。
2. 蛋液半個、生粉 2 湯匙拌成蛋漿，備用。
3. 將齋餡料（約 2 湯匙）鋪在小塊腐皮上，包捲成長方形（7cm X 3cm），用蛋漿塗在邊緣密封。
4. 用中火燒熱鑊，下油滑鑊，將油盛起，轉小火後放入腐皮卷，慢火煎至兩面呈金黃色及香脆，盛起，吸去油分，以喼汁或香醋蘸吃。

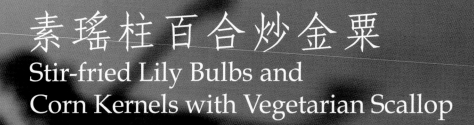

素瑤柱百合炒金粟
Stir-fried Lily Bulbs and
Corn Kernels with Vegetarian Scallop

材料：		調味料：	
金菇	2 紮	清水	半杯
鮮百合	150 克	鹽	半茶匙
西芹	150 克	素雞粉	半茶匙
粟米粒	150 克	砂糖	半茶匙
甘筍	120 克	油	1 茶匙

芡汁：

鹽	半茶匙
素雞粉	1/3 茶匙
素蠔油	1 茶匙
生粉	1 茶匙
清水	3 茶匙

購買食材 tips

- 鮮百合一般是真空包裝，這菜譜的份量約用 2 包。最好挑選表面沒呈褐色枯謝的。
- 粟米粒在凍肉店或超市購買急凍包裝貨品，也可購買包裝雜豆（有粟米粒、甘筍粒及青豆），或買原條新鮮粟米蒸熟後取粒使用。

前一天準備篇

· 鮮百合拆散、洗淨，用小刀削去枯謝部分，冷藏備用。

· 若用原條粟米，可預先蒸熟後，剝出原粒，冷藏留用。

1. 金菇切去尾段棄掉，再切成約半吋長小段，拆散，洗淨，瀝乾水分。

2. 西芹削去表皮，洗淨，與甘筍一起切粒（約半吋丁方），備用。

3. 燒熱鑊，下生油用中火油溫，放入金菇炸至金黃色，瀝乾油分。

4. 鑊內放入清水半杯，下調味料用大火燒滾，放入所有蔬菜料煸炒約2分鐘，盛起備用。

5. 鑊燒熱下生油2茶匙，下蔬菜粒炒勻，加入已調勻的芡汁拌炒，上碟，灑上炸金菇（炸瑤柱）即成。

> 「西芹先用刨皮器去掉外皮，再用手撕去硬筋。」

> 「煸炒是用少量湯或汁，短時間炒熟材料，特別適用於蔬菜類。」

> 「金菇必須鬆散入鑊，炸後如同炸瑤柱。」

椰香南瓜盅
Pumpkin Bowl with Coconut Milk

材料：

日本南瓜	1 個
鮮草菇	150 克
蘑菇	10 粒
馬蹄肉	10 粒
甘筍	100 克
素蝦仁	1 包
芋頭	300 克
齋湯	2/3 杯（參考 p.148）
椰奶	1/5 杯

* 草菇及馬蹄肉選用齋湯材料，
  參考 p.148。

調味料：

鹽	半茶匙
素雞粉	1 茶匙
糖	半茶匙

購買食材 tips

- 選購瓜身呈圓形的日本南瓜，挑選瓜身重、外形完好、色澤深橙紅色、表皮堅硬及莖蒂較乾的（表示成熟及甜味充足），可預早購買放置陰涼處，可保存 1 至 2 個月。
- 可用新鮮蘑菇，較罐裝的更鮮味。
- 素蝦仁於健康素食店有售，大部分貨品是真空包裝。
- 芋頭最宜選購較乾身的為佳，避免有生水芋頭情況出現。
- 椰奶選用罐裝椰奶。

前一天準備篇

南瓜盅做法：

1. 南瓜洗淨，在頂部約 1 吋厚度橫切做成瓜蓋，去籽及內瓢，洗淨內部。

2. 用鑊隔水蒸約 8 分鐘至 5 成腍，取出待涼，冷卻後用保鮮紙密封，放入雪櫃貯存。

當天宴客篇

1. 南瓜盅在雪櫃取出，在室溫解凍。

2. 蘑菇、馬蹄一開二切開。

3. 芋頭去皮，切成2cm立方；甘筍切粗粒。

4. 鑊內放入清水2杯煮滾，下草菇、蘑菇、甘筍、馬蹄肉、素蝦仁，加鹽1茶匙及薑汁2茶匙滾透約3分鐘，盛起瀝水備用。

5. 鑊燒熱下生油約2茶匙，中火將芋頭煎香（約2分鐘），加入齋湯2/3杯，再放入所有材料煮滾，轉用慢火爛至芋頭表面開始溶化，加入調味料煮3分鐘後，放入椰奶拌勻，轉放南瓜盅內。

6. 南瓜盅隔水蒸約12分鐘，取出食用。

素丁雙色炒飯
Stir-fried Two-colour Rice
with Diced Vegetables

雙色飯材料：		炒飯材料：	
白米	2 1/2 量米杯	素餐肉	200 克
紅糙米	半量米杯	菜心	100 克
清水	3 杯	甘筍	100 克
		冬菇	5 朵

調味料：		（選用齋湯材料，參考 p.148）	
鹽	1 茶匙	松子仁	50 克
素雞粉	1 茶匙	雞蛋	2 個
生抽	2 茶匙		

購買食材 tips

- 素餐肉在健康素食店有售，多是真空包裝，有片或原件包裝，買原件裝可切配成粒狀。
- 松子仁有散裝生松子仁，或於超市買已烘脆真空包裝的，可省卻烘製松子仁的工序。

前一天準備篇

雙色飯做法：

1. 紅糙米洗淨，<u>用 1 杯清水浸泡約 3 小時</u>備用。
2. 白米洗淨後，與紅糙米（連浸泡米水）同放入飯煲內，加清水 2 杯煲熟，待涼後冷藏。

> 「省時的話，紅糙米用溫水浸 2 小時即可。」

松子仁做法：

鑊燒熱，用白鑊慢火炒松子仁至表面呈淡金黃色（約 4 至 5 分鐘），盛起待涼，存放密封瓶或盒內，避免受空氣潮濕影響。

165

當天宴客篇

1. 將雙色飯預早在雪櫃取出，或<u>在微波爐翻熱</u>備用。
2. 將所有材料切粒洗淨，用滾水灼熟備用。
3. 雞蛋拂成蛋液，鑊燒熱後下生油 2 茶匙，轉中火下蛋液炒成半熟。
4. 放入雙色飯快炒數下拌勻，再轉中小火炒熱，加入材料同炒至熱，下調味料炒約 1 分鐘，最後灑上松子仁炒勻，即可上碟。

> 「家裏沒微波爐，可蒸熱使用。」

花旗參桂花糕
Sweet Osmanthus Pudding
with American Ginseng

材料：

花旗參片	5 克
杞子	20 克
糖桂花	1 茶匙
桂花糖水	3 湯匙（糖桂花內之糖水）
黃砂糖	60 克
清水	2 杯
魚膠粉	20 克

購買食材 tips

- 糖桂花在雜貨店有售，九龍城街市的潮州食材店或雜貨店均有，糖水與桂花混合浸着，一般以玻璃小瓶裝出售，每樽約 20 至 25 元。
- 花旗參片及杞子在藥材店有售。

前一天準備篇

花旗參桂花糕做法：

1. 魚膠粉加入清水 100ml 浸約 5 分鐘，或見魚膠粉吸收至發漲，備用。

2. 杞子用清水浸軟，隔水備用。

3. 將清水 2 杯煮滾，熄火，下花旗參片浸焗約半小時。

4. 再將花旗參水煮熱，加入桂花糖水、糖桂花、黃砂糖及杞子拌勻。

5. 魚膠粉放入微波爐，用中火翻熱 10 秒，加入上述糖水內拌勻，傾入糕模或窩碟，待涼後冷藏至凝固。

> 「魚膠粉與水的比例一般為 1 比 5。」

> 「花旗參片宜浸焗，因過份煲煮後，其性質變為較溫熱。」

當天宴客篇

將冷藏的桂花糕取出，切件即可品嘗。

大廚小宴　FEAST of Master Chef

作者	Author
廖教賢	Alvin Liu
策劃/編輯	Project Editor
	Karen Kan
攝影	Photographer
	Imagine Union
美術統籌	Art Direction
	Ami
美術設計	Design
	Man
封面設計	Cover Design
	Mandi Leung

出版者　Publisher
Forms Kitchen
香港鰂魚涌英皇道1065號東達中心1305室
Room 1305, Eastern Centre, 1065 King's Road, Quarry Bay, Hong Kong
電話　Tel:　2564 7511
傳真　Fax:　2565 5539
電郵　Email: info@wanlibk.com
網址　Web Site:　http://www.wanlibk.com
　　　　　　　　http://www.facebook.com/wanlibk

發行者　Distributor
香港聯合書刊物流有限公司
SUP Publishing Logistics (HK) Ltd.
香港新界大埔汀麗路36號
3/F, C&C Building, 36 Ting Lai Road,
中華商務印刷大廈3字樓
Tai Po, N.T., Hong Kong
電話　Tel:　2150 2100
傳真　Fax:　2407 3062
電郵　Email: info@suplogistics.com.hk

承印者　Printer
中華商務彩色印刷有限公司
C & C Offset Printing Co., Ltd.

出版日期　Publishing Date
二〇一三年七月第一次印刷　First print in July 2013
二〇一八年四月第三次印刷　Third print in April 2018

Crisp Beancurd Skin Rolls

Tips for shopping

- Buy thin, round, translucent beancurd skin (about 2 1/2 feet in diameter). It is generally available at soy food shops at about $4 to $5 each. Some shops offer packed beancurd skin in two sheets. It should be kept refrigerated in an airtight plastic bag at home to avoid drying and rotting.
- Choose pork collar-butt which is evenly spread with fat and lean meat. It is better than just lean meat.

Ingredients

2 sheets beancurd skin
100 g minced pork

Seasoning

5 tbsps water
1 tbsp oyster sauce
2 tsps caltrop starch
ground white pepper
sesame oil

169

One day before feast

Mince the pork by chopping or blending until it is smooth. Mix in the seasoning. Keep in a refrigerator overnight.

On the day of feast

1. Cut the beancurd skin in the middle into two semi-circular sheets. Spread the seasoned minced pork evenly on the beancurd skin. Roll into a cylinder. Cut into small rolls of about 2 cm wide.
2. Put about 2 cups of oil in a wok over medium heat. When it is hot, deep-fry the beancurd skin rolls until golden. Drain. Put on a plate. Serve with Worcestershire sauce.

Fish Maw and Tofu Thick Soup

Tips for shopping

- Dried fish maw is divided into two types: deep-fried fish maw is made by deep-frying in oil. The product is yellowish and easily smells odd and oily. It is cheaper but cannot be kept long. Another type is fried fish maw, which is made by stir-frying in hot sand. The product is whiter and dry. It has a better performance in rehydration and is more expensive.
- Vacuum-packed Jinhua ham can be used. Its weight is about 75g to 113 g. Although it is cheaper, its quality is not as good as that of a whole ham.
- You can buy chicken stock in cans or boxes.

Ingredients

150 g dried fish maw
2 pieces cloth-wrapped tofu
10 g Jinhua ham
500 ml chicken stock
2 egg whites
500 ml water
4 slices ginger
2 sprigs spring onion
1 tbsp rice vinegar

Seasoning

2/3 tsp salt
1/2 tsp sesame oil
ground white pepper

Caltrop starch solution

80 ml water
30 g caltrop starch
* mixed well

One day before feast

Method for dried fish maw:

1. Rehydrate the dried fish maw by soaking in cold water (about 2 hours). Cut into dices.
2. Heat water in a wok. Add the rice vinegar, ginger and spring onion. Bring to the boil. Put in the fish maw. Boil for about 3 minutes to remove the odd smell. Dish up. Cool in cold water and drain. Keep refrigerated.

Method for Jinhua ham:

1. Slightly blanch the Jinhua ham (about 1 minute). Put into a bowl. Add water to cover the ham. Steam for about half an hour. When it cools, dish up.
2. Cut the Jinhua ham into squares of 2 inches long (reserved for making "Winter Melon and Roast Duck Rolls"). Finely chop the rest of the ham and keep refrigerated. Add the steamed ham water into the chicken stock.

On the day of feast

1. Dice the cloth-wrapped tofu. Soak in boiling water until hot. Set aside.
2. Take out the fish maw from the refrigerator. Boil thoroughly. Drain.
3. Put the water and chicken stock in a pot. Heat up. Add the fish maw. Drain the tofu and put into the soup. Add the seasoning and bring to the boil. Stir the soup. Mix in the caltrop starch solution. When it comes to the boil, turn off heat. Mix in the egg whites. Put into a bowl. Sprinkle with the chopped Jinhua ham. Serve.

Winter Melon and Roast Duck Rolls

Tips for shopping

- Better select a heavy, fleshy winter melon with smooth surface and no cracks.
- You can buy the vacuum-packed and boned Jinhua ham.
- It is the best to pick the meaty part of the roast duck, such as the leg or breast.

Ingredients

1 1/2 kg winter melon
1 roast duck leg
5 dried black mushrooms
200 g carrot
60 g Jinhua ham
300 g Chinese chives
600 g flowering Chinese cabbage
(about 16 stalks)

Thickening sauce

80 ml water
1 tbsp oyster sauce
1/3 tsp salt
1/2 tsp chicken bouillon powder

Caltrop starch solution

2 tbsps caltrop starch
5 tbsps water
* mixed well

One day before feast

- Peel the winter melon. Remove the seeds. Cut into a few rectangles measuring 5 inches x 3 inches. Cut straight into about 20 slices, each around 2mm thick. Seal with cling wrap. Keep in a refrigerator.
- Refer to the method of dried black mushrooms on p.10. When it cools, cut into coarse strips. Keep refrigerated.
- Take the pieces of Jinhua ham from the "Fish Maw and Beancurd Thick Soup". Cut into strips (about 2 inches long and 1 cm wide). Keep refrigerated.
- Cut the carrot into strips same as the shape of Jinhua ham.

On the day of feast

1. Soak the winter melon slices and Chinese chives in boiling water for about 1 minute. Dish up. Cool in cold water and drain.
2. Bone the roast duck leg. Cut into strips (about 2 inches long). Set aside.
3. Lay a slice of winter melon flat. Put 1 strip each of the roast duck, carrot, Jinhua ham and black mushroom on top. Roll up. Fasten with the Chinese chive. Make around 20 rolls.
4. Put the roast duck rolls on a plate. Steam over high heat for about 6 minutes. Take out. Blanch the flowering Chinese cabbage until done. Drain. Put on the side of the plate.
5. Heat the thickening sauce. Add the caltrop starch solution. Bring to the boil. Sprinkle on the roast duck rolls. Serve.

171

Stir-fried Asparagus and Shrimp Slices

Tips for shopping

- Do not pick fresh asparagus which is too thick or too thin. Asparagus from Australia is the best. Although those from mainland China or Thailand are cheaper, their quality is not promising.
- Frozen shrimp meat can be used. Packed shrimp meat of 1 kg in weight is available at supermarkets or frozen meat shops. It is generally produced in Vietnam or Thailand.

Ingredients

800 g fresh asparagus
400 g shelled shrimps
100 g skinned water chestnuts
 (about 6 to 7 pieces; chopped up)

Spices

finely chopped garlic
ginger slices
flower-shaped carrot slices

Seasoning for minced shrimp

2/3 tsp salt
1 tsp chicken bouillon powder
2 tbsps caltrop starch
1 tsp sesame oil
1/3 tsp ground white pepper
1/2 egg white

Seasoning

1/3 tsp salt
1/2 tsp chicken bouillon powder
1 tsp caltrop starch
3 tsps water
1/2 tsp sesame oil
* mixed well

One day before feast

Method for minced shrimp:

1. Defrost the shrimp meat (if fresh shrimp is used, remove the shell). Rub the shrimp meat with caltrop starch. Rinse well.
2. Sop up the water in the shrimp meat with dry cloth. Pat the shrimp meat with a knife. Chop with the back of the knife.
3. Put the shrimp meat into a big bowl. Knead by hand for about 3 minutes, or until it is a bit sticky. Add the seasoning for minced shrimp. Stir in one direction (about 5 minutes) until it is sticky and spongy. Throw the shrimp meat into the bowl by hand repeatedly until it is gluey. Mix in the shelled water chestnuts. Keep refrigerated.

On the day of feast

1. Cut away the old strings at the end of the asparagus (about 4 cm). Scrape off the hard skin. Rinse and cut diagonally into thick slices.
2. Heat a wok. Put in 1 tbsp of oil. Add the minced shrimp and flatten with a turner. Fry over low heat until both sides are golden and done. Drain. Cut diagonally into slices.
3. Bring 1/2 cup of water in a wok to the boil. Add 1 tbsp of ginger juice with wine, and 1 tsp each of salt, sugar, and oil. Put in the asparagus. Stir-fry over medium heat until 70% done (about 1 minute). Drain.
4. Heat a wok. Add 1 tsp of oil. Stir-fry the spices over low heat until fragrant. Put in the asparagus and shrimp slices. Swiftly stir-fry over medium heat. Pour in the seasoning. Give a good stir-fry. Sprinkle wine along the edge of the wok. Swiftly stir-fry. Dish up and serve.

Deep-fried Pork in Spiced Orange Sauce

Tips for shopping

- Choose fresh pork rib eye or chilled boneless pork chop.
- Buy Japanese coarse bread crumbs which are fluffier and crisper.
- Sunquick concentrated orange juice can be used. It has a rich orange flavour.

Ingredients

500 g pork rib eye
1 orange
150 g bread crumbs
1 egg

Marinade

1/2 tsp salt
1/2 tsp chicken bouillon powder
4 tsps caltrop starch
1 egg
4 tsps concentrated orange juice
4 tbsps water

Spiced orange sauce

50 ml white vinegar
100 ml water
100 ml concentrated orange juice
60 g sugar
1 orange
(peeled and diced; added last)
2 tsps orange liqueur or gin (added last)
20 g custard powder

One day before feast

Method for pork:

Cut the veins away from the pork eye rib. Cut into thick slices (about 0.5 cm thick). Rinse and drain. Mix the marinade. Put in the pork eye rib and mix well. Keep refrigerated.

Method for spiced orange sauce:

Cook the white vinegar and orange juice over low heat. When it is hot, add the sugar and cook until it dissolves. Thicken the sauce with the custard powder mixed with water. Finally mix in the orange liqueur and diced orange. When it cools, keep refrigerated.

On the day of feast

1. Coat the marinated pork with bread crumbs. Press tightly by hand to keep the bread crumbs intact.
2. Heat 2 cups of oil in a wok. Deep-fry the pork over medium heat until golden and done. Sop up the oil. Put on a plate. Heat the spiced orange sauce. Sprinkle on the pork. Serve.

173

Salt-baked Grey Mullet with Lemongrass

Tips for shopping

The grey mullet commonly found in the market is chilled. Select those that are plump with red gills and intact scales.

Ingredients

2 grey mullets (about 600 g)
4 stalks lemongrass
6 sprigs coriander
4 sprigs spring onion
1.2 kg coarse salt
4 lemon leaves

Dipping sauce

3 tbsps fish sauce
3 tbsps boiled water
1 tbsp sugar
2 limes (juice squeezed)
1 tsp white rice vinegar
3 sprigs coriander stems (chopped up)
2 bird's eye chillies (diced)
1 tsp finely chopped garlic
* stir until the sugar dissolves

One day before feast

It is not necessary to prepare seafood in advance.

On the day of feast

1. Preheat an oven to 220℃ for 10 minutes.
2. It is no need to scale the fish. Just rinse and wipe dry.
3. Section the lemongrass. Put 2 stalks of lemongrass, 3 sprigs of coriander, 2 sprigs of spring onion and 2 lemon leaves into the cavity of each fish.
4. Lay aluminum foil on a baking tray. Spread the coarse salt on the aluminum foil. Put the fish on top. Cover the fish with the coarse salt. Wrap up and seal the fish. Put in the oven and bake for about 25 minutes. Take out.
5. When ready to eat, brush away the salt. Remove the spices from the fish cavity. Flatten the fish cavity. Put on a plate. Serve with the dipping sauce.

Dumplings and Noodles in Soup

Tips for shopping

- Guangdong raw noodles (noodles with a soapy texture), divided into thick and thin types, are used by noodle shops in serving with wontons. Loosen the noodles and keep them in a bag at home. It can help remove their soapy flavour.
- Guangdong and Shanghai dumpling wrappers are quite different. The former is thin and yellowish while the latter is thick and white. They can be bought at shops selling raw noodles. Keep the dumpling wrappers in an airtight container to avoid drying.
- Picking hairy wood ear fungus, which is large in size with a crunchy texture.
- You can buy stock in cans or boxes.

Ingredients

4 pieces Guangdong raw noodles
300 g Guangdong dumpling wrappers (about 25 sheets)
450 g flowering Chinese cabbage (about 12 stalks)
500 ml stock
300 ml water
30 g yellow chives

Stuffing

20 fresh shelled shrimps
400 g pork collar-butt
1 piece wood ear fungus
40 g dried shrimps
40 g ginger

Seasoning

1/2 tsp salt
1 tbsp oyster sauce
1 tsp chicken bouillon powder
4 tsps caltrop starch
2 tbsps water
1 egg white
1 tsp sesame oil
1/3 tsp ground white pepper

One day before feast

Method for stuffing:

1. Rub the shelled shrimps with caltrop starch. Rinse well. Sop up the water with dry cloth. Coarsely dice. Set aside.
2. Soak the wood ear fungus in water to soften. Finely shred. Slightly blanch to remove the odd smell. Cool in cold water.
3. Soak the dried shrimps in water to soften (about 15 minutes). Rinse and drain. Finely chop. Stir-fry in a dry wok until fragrant. Set aside.
4. Finely chop the ginger. Rinse and chop up the pork collar-butt.
5. Put all the ingredients into a bowl. Add the seasoning. Stir until gluey. Keep refrigerated.

175

On the day of feast

1. Wrap the stuffing in the dumpling wrapper (about 25 pieces).
2. Trim the flowering Chinese cabbage. Take the tender parts and rinse. Dice and rinse the yellow chives.
3. Blanch the raw noodles until done (do not overcook to keep them crunchy). Dish up. Rinse in cold water. Put into boiling water again to heat the noodles. Transfer to a big bowl.
4. Blanch the dumplings and flowering Chinese cabbage separately until done. Transfer to the big bowl.
5. Bring the stock and water to the boil (if it is not salty enough, season with 1/3 tsp of salt). Sprinkle with the yellow chives. Pour the soup into the bowl. Serve.

Cold Green Tea Pudding

Tips for shopping

- Green tea powder, generally from Taiwan, China and Japan, is available at supermarkets. The product from Japan has a richer flavour. For packing, it has a choice of small cans and bags (containing small packets of 10 g to 15 g each). It is better to buy green tea powder in bags for better storage of the unused ones.
- Regular fresh milk in boxes can be used. If cream is used, reduce the quantity to 50 ml as it has a richer milky flavour and is creamier than regular milk.

Ingredients

50 g green tea powder
300 ml fresh milk
1 cup water
70 g sugar
26 g gelatin powder

One day before feast

- Cover the gelatin powder with 1/2 cup of boiled water. Stir evenly to let it absorb the water.
- Mix the green tea powder with fresh milk until it dissolves. Strain with a mesh strainer to remove any lumps.
- Bring water to the boil. Add the sugar. When it dissolves, stir in the green tea milk.
- Put the gelatin powder in a microwave oven. Turn to medium heat. Heat for 30 seconds until dissolves (or sit the gelatin powder on hot water to dissolve). Put into the hot green tea milk and mix well. Pour into a dessert cup or cake tin. Let it cool down. Chill to set.

On the day of feast

Take out the cold green tea pudding from the refrigerator. Serve with fresh fruit, evaporated milk or cream.

Spicy Garlic Duck Tongues

Tips for shopping

- Duck tongues are frozen products sold at regular frozen food shops in packs.
- Bottled ground Sichuan peppercorn and Sichuan peppercorn oil are available at large supermarkets and groceries.

Ingredients

600 g frozen duck tongues
30 g diced red chillies
2 tsps white sesame seeds

Ingredients of Chinese marinade

2 cups water
1/3 cup light soy sauce
2 tbsps oyster sauce
1 tsp salt
1 piece slab sugar (about 60 g)
2 tsps chicken bouillon powder
1 tbsp ground Sichuan peppercorn
40 g spices for Chinese marinade
30 g finely chopped garlic
20 g diced ginger
20 g diced shallot
2 tbsps oil
2 tbsps Shaoxing Hua Diao wine
 (added last)

Seasoning for duck tongues

40 g finely chopped garlic
1 tsp ground Sichuan peppercorn
1 tsp Sichuan peppercorn oil
2 tsps sesame oil
2 tsps chilli bean sauce
1 tbsp oyster sauce
2 tbsps sugar

One day before feast

Method for spiced duck tongues:

1. Stir-fry the garlic, ginger and shallot of the Chinese marinade ingredients with oil until fragrant. Add the rest Chinese marinade ingredients. Bring to the boil over medium heat. Turn to low heat and simmer for about 20 minutes. Turn off heat. Leave with a lid on. When it cools and the Chinese marinade is flavourful, remove the spices. Keep the sauce.
2. Defrost the duck tongues and rinse. Bring water in a pot to the boil. Put in the duck tongues. Bring to the boil. Turn off heat. Leave with a lid on for 10 minutes. Remove the duck tongues. Cool in cold water. Drain.
3. Put the duck tongues in the Chinese marinade. Keep refrigerated for about 3 hours. Dish up the duck tongues. Keep refrigerated again.

177

On the day of feast

1. Take 20 g of chopped garlic of the seasoning for duck tongues. Stir-fry with 2 tbsps of oil over low heat until fragrant. Set aside.
2. Combine the seasoning for duck tongues together. Add the fried garlic. Mix well.
3. Stir-fry the white sesame seeds in a dry wok over low heat until fragrant and golden. Let it cool down.
4. Take out the spiced duck tongues. Mix with the seasoning for duck tongues. Put on a plate. Finally sprinkle the white sesame seeds and red chillies on top.

Wontons and Chicken in Casserole

Tips for shopping

- Either fresh or chilled chicken can be used.
- You can pick vacuum-packed Jinhua ham meat in the market. As for the bone, it is most suitable to buy the leg part with skin on, which smells fragrant. The best, of course, is to buy the parts cut from a whole Jinhua ham.
- Choose Shanghai wonton wrappers instead of Guangdong ones. It is thick and does not crumble easily.
- Use Peking cabbage which is thick and crunchy.

Ingredients

1 chicken
200 g Jinhua ham meat
500 g Jinhua ham bone
400 g pork spareribs
600 g baby cabbage
4 big slices ginger
14 cups water

Ingredients of pork and vegetable wontons

300 g pork collar-butt
300 g Peking cabbage
40 g diced ginger
24 Shanghai wonton wrappers

Seasoning for chicken stock

1 tsp salt

Seasoning for wontons

1/2 tsp salt
1/2 tsp chicken bouillon powder
1 tbsp oyster sauce
1 egg white
2 tsps caltrop starch
3 tsps water
ground white pepper
sesame oil

One day before feast

Method for pork and vegetable wontons:

1. Rinse the pork collar-butt. Mince by blending or chopping.
2. Rinse and chop up the Peking cabbage. Slightly blanch. Cool in cold water and drain. Squeeze out water.
3. Mix the Peking cabbage, ginger and seasoning with the minced pork. Stir until sticky to be the stuffing. Cover with cling wrap. Keep refrigerated overnight.

On the day of feast

1. Wash the chicken and pork spareribs. Slightly blanch the Jinhua ham and spareribs to remove grease and blood.
2. Put 3.5 kg of water into a large pot. Bring to the boil. Add the spareribs, ham and ginger. Cook over high heat for half an hour. Turn to low-medium heat and cook for about 1 1/2 hours as soup base. Put in the chicken and cook over medium heat for 1 1/2 hours.
3. Cut the baby cabbage into long strips and rinse.
4. Take out the stuffing. Wrap in the Shanghai wonton wrappers. Set aside.
5. Remove the spareribs from the soup. Take the meat and dump the bone. Put the chicken and the meat in a casserole. Tear the ham into shreds. Pour the soup with the shredded ham into the casserole. Cook over low-medium heat until it comes to the boil.
6. Blanch the wontons and baby cabbage until done. Transfer to the casserole. Bring the soup to the boil over low heat (about 10 minutes). Serve hot.

Steamed Loin Ribs with Glutinous Rice on Lotus Leaf

Tips for shopping

- Lotus leaf is divided into the fresh and dried. Available in summer, fresh lotus leaf is bright in colour and smells a light aroma. Dried lotus leaf looks withered but its fragrance is intense. It can be found in regular markets.
- Loin ribs have an even distribution of fat and lean meat. You may choose pork spareribs which have more lean meat.

Ingredients

600 g loin ribs
300 g glutinous rice
20 g Qi Zi
1 dried lotus leaf
1 sprig coriander

Spices

10 g finely diced ginger
20 g finely chopped garlic
1 small piece dried tangerine peel
(finely diced)

Marinade

1 tsp chilli bean sauce
2 tbsps oyster sauce
1/2 tsp salt
1 tsp sugar
1/2 tsp five-spice powder
1 tbsp cooking wine (Hua Diao or double distilled rice wine)
2 tbsps caltrop starch
1/4 tsp ground white pepper
1 tsp sesame oil
4 tbsps water

One day before feast

- Chop the loin ribs into rectangles of 1 inch x 2 inches. Rinse and drain. Mix the marinade with the spices. Add the loin ribs and mix well. Keep refrigerated.
- Buy the dried lotus leaf in advance. Put into an airtight plastic bag and keep refrigerated.

On the day of feast

1. Wash the glutinous rice. Cover with water and soak for about 3 to 4 hours.
2. Soak Qi Zi in water until soft. Rinse.
3. Slightly blanch the dried lotus leaf. Cool in cold water and drain. Lay on a steamer or plate.
4. Drain the glutinous rice. Put into a big bowl. Put in the loin ribs and fully coat with glutinous rice. Arrange on the lotus leaf. Sprinkle with Qi Zi. Steam for about 25 minutes. Finally decorate with the coriander. Serve.

179

Stir-fried Pork and Vegetable Shreds with Chrysanthemum

Tips for shopping

- Rib eye is the lean meat part of the pork. Pork spareribs can also be used. Bone the spareribs and then shred the meat.
- Choose cheaper black mushrooms for shredding. It is no need to pick the thick ones.
- Either yellow or white chrysanthemum can be used. You can buy it with the lemon leaf in flower shops.
- Pick corn flakes in original flavour, which are lighter in taste. They are laid on the bottom of the dish, and will not affect the flavour of the dish.

Ingredients

200 g pork rib eye
200 g roast pork (finely shredded)
10 dried black mushrooms
 (refer to p.10 for method; shredded)
300 g celery
100 g carrot
1 yellow chrysanthemum
1 lemon leaf
60 g corn flakes in original flavour

Spices

1/3 tsp finely chopped garlic
10 g shredded ginger
20 g shredded green bell pepper
20 g shredded red bell pepper

Marinade

2 tsps light soy sauce
1/2 tsp sugar
2 tsps caltrop starch
2 tbsps water

Seasoning for blanching

2 cups boiling water
1 tsp salt
1/2 tsp sugar
1 tsp oil

Seasoning

1/2 tsp salt
1/2 tsp chicken bouillon powder
1 tsp oyster sauce
1 tsp dark soy sauce
3 tbsps water
2 tsps caltrop starch

One day before feast

Slice the rib eye into 0.5 cm thick. Flatten and cut into shreds. Rinse and drain. Mix the marinade together. Put in the shredded pork and mix well. Keep refrigerated.

On the day of feast

1. Shave off the hard strings of the celery. Rinse and finely shred.
2. Peel the carrot and coarsely slice. Cut into thin strips.
3. Rinse and finely shred the lemon leaf. Cut away the receptacle of the chrysanthemum. Soak the petals in salt water for a couple of minutes. Drain.
4. Bring 2 cups of water to the boil. Add the seasoning for blanching. Bring to the boil over high heat. Add the celery and carrot. Boil for about 2 minutes. Drain.
5. Heat a wok. Add 2 tsps of oil. Stir-fry the shredded pork over medium heat until done. Set aside.
6. Heat the wok again. Add 2 tsps of oil. Stir-fry the spices until fragrant. Put in the mushrooms, celery and carrot. Stir-fry over medium heat for a while. Add the roast pork and shredded pork. Give a good stir-fry.
7. Mix the seasoning. Pour into the wok bit by bit. Stir-fry swiftly over high heat. Sprinkle with wine and stir-fry swiftly.
8. Lay the corn flakes on a plate. Arrange the shredded pork and vegetables on top. Finally sprinkle with the chrysanthemum petals and lemon leaf shreds. Serve.

Honey Glazed Roast Pork with Baguette

Tips for shopping

- Either fresh or frozen pork collar-butt can be used. The fresh one is chewier. The frozen one, after baked, has a soft and fluffy texture. It is available at frozen food shops with a general packing of over 2 kg.
- You can choose the multi-flora honey prepared for barbecue.

Ingredients

1 kg pork collar-butt
1 bottle multi-flora honey
120 g caltrop starch

Ingredients of garlic baguette

1 baguette
30 g finely chopped garlic
40 g salted butter

Marinade

2 tsps salt
2 tsps chicken bouillon powder
5 tbsps sugar
2 tbsps oyster sauce
1 tsp light soy sauce
1 tsp dark soy sauce
3 tbsps Hoi Sin sauce
1 tbsp sesame sauce
2 tsps Chinese rose wine
1 tsp finely chopped shallot
1 tsp finely chopped garlic
1 egg

Honey glaze

2 tbsps Hoi Sin sauce
3 tbsps multi-flora honey
1 tsp light soy sauce
1 tbsp boiling water
30 g finely diced onion

One day before feast

- Mix the marinade. Seal with cling wrap and refrigerated (if the weather is cool, it may be kept in room temperature).
- Cut the pork collar-butt in long strips of about 8 inches long x 2.5 inches wide x 1 inch thick. Rinse well. Mix with 60 g of caltrop starch. Seal and keep refrigerated overnight.
- Heat a wok. Add 1 tsp of oil. Stir-fry the onion until fragrant. Put in the honey glaze and heat up. Let it cool down. Keep refrigerated.

On the day of feast

1. Take out the pork collar-butt. Wash away the caltrop starch and blood. Mix in 60 g of caltrop starch. Mix with the marinade and rest for 40 minutes.
2. Turn up the heat (about 240℃) of an oven. Preheat for 15 minutes. Lay aluminum foil on the bottom tray to avoid spotting.
3. Lay the pork collar-butt evenly on the oven rack. Bake over high heat for about 10 minutes. Flip over and bake the other side for 10 minutes, or until the surface is reddish and the edges are slightly burnt. Turn down the heat (about 120 ℃) and bake for 30 minutes. Take out. Spread the honey on the surface.
4. Turn up the heat (about 250℃ to 280℃) of the oven. Bake each side of the roast pork for about 5 minutes. Turn down the heat and bake for 30 minutes. Take out. Spread with the honey again. Let it cool down.
5. During the time of baking, cut across the baguette into around 1 cm thick pieces. Mix the garlic with butter. Spread on the surface of the baguette.
6. When the roast pork is done, take out. Put in the baguette and bake over low heat for about 5 minutes.
7. Put the honey glaze in a microwave oven. Heat up. Set aside.
8. When the baguette is baked crunchy, take out and put on a plate. Slice the roast pork diagonally (about 1 cm thick). Put on top of the baguette. Spread with the honey glaze. Serve.

181

Cheese Stuffed Shrimp Balls

Tips for shopping

- Buy frozen shrimp meat which is handy and cheap. If fresh shrimp is used, remove the shell and rub with caltrop starch to clean the surface. Wash and drain. It needs to be chilled for a couple of hours before mincing.
- It is convenient to use Kraft cheese slices. It is not so salty. If choosing other cheese, pay attention to the degree of its saltiness.
- Use Japanese bread crumbs as coating for deep-frying. It gives a crunchier texture with a golden tint.

Ingredients

500 g shelled shrimps
10 skinned water chestnuts (120 g)
3 cups oil
250 g bread crumbs

Cheese stuffing

4 slices Kraft cheese
30 g butter
20 g crab roe
50 ml water
2 tsps sugar

Seasoning

1 tsp salt
1/2 tsp chicken bouillon powder
1/2 tsp sugar
1/2 tsp sesame oil
1/4 tsp ground white pepper
2 tbsps caltrop starch (added last)
1/2 egg white (added last)

One day before feast

Method for cheese stuffing:

Put the cheese slices, butter and water into a bowl. Steam to melt. Mix in the sugar and crab roe. Pour into a container. Let it cool down. Put into a refrigerator. When it is set, cut into small dices. The cheese stuffing is done.

Method for minced shrimp:

1. Rinse the skinned water chestnuts. Pat with a knife and chop up. Press out water.
2. Defrost the shelled shrimps and rinse. Sop up the water with dry cloth or paper towel. Pat the shrimps with the knife. Finely chop with the back of knife. Put into a big bowl. Add the seasoning. Stir in one direction until sticky. Add the caltrop starch and stir until gluey. Put in the egg white and stir evenly. Knead and throw repeatedly into the bowl until it is very sticky. Chill overnight.

On the day of feast

1. Pump the minced shrimp into small balls of about 1 inch in diameter. Stuff with the small cheese dices. Gently knead into round shape. Coat entirely with the bread crumbs.
2. Put oil in a wok or small pot (the oil should be enough to cover the shrimp balls). Heat the oil until medium hot (about 200 ℃). Put in the shrimp balls (if there is insufficient oil, put in the shrimp balls separately for a couple of times). Deep-fry for about 1 to 2 minutes. Turn down the heat. Deep-fry until the shrimp balls float and the surface is slightly golden. Turn to medium heat and deep-fry for about 2 minutes. Drain. Put on a plate. Serve with salad dressing.

Fried Rice with Salted Fish, Dried Prawns and Vegetables

Tips for shopping

- Use the Mei Xiang type salted fish which is more fragrant and flavourful. For the sake of convenience, you can buy bottled salted fish paste in regular supermarkets.
- Pick flowering Chinese cabbage with long, thick stems. It is crunchy. Chinese celery, having a strong aroma, can be used instead.
- Select dry, golden and fragrant dried prawns. Dried marine prawns taste fresh and sweet, but they are more expensive.

Ingredients

1 piece salted fish (about 60 g)
80 g dried prawns
10 stalks flowering Chinese cabbage (about 400 g)
1 tbsp diced spring onion
1/2 tbsp diced ginger
600 g steamed rice
2 eggs

Seasoning

1 tsp salt
1/2 tsp chicken bouillon powder

One day before feast

- Fry the salted fish over low heat until fragrant. Finely dice. Keep refrigerated.
- Soak the dried prawns in water to soften. Rinse and finely dice. Keep refrigerated.

On the day of feast

1. Rinse the flowering Chinese cabbage. Finely slice.
2. Heat a wok. Put in 2 tsps of oil. Add the ginger, vegetables, salted fish and dried prawns. Stir-fry over medium heat until aromatic. Pour in the whisked eggs. Stir-fry for a while.
3. Add the steamed rice. Turn to low heat. Stir-fry the egg wash evenly. Adjust to medium heat. Swiftly stir-fry until the steamed rice is hot through. Sprinkle with the seasoning. Stir-fry evenly until the seasoning dissolves. Finally sprinkle with the spring onion. Turn to high heat and swiftly stir-fry for a moment. Put on a plate. Serve.

183

Taro Sago Sweet Soup with Coconut Cream

Tips for shopping

- Better buy Li Pu taro (from China) with dry, plump body and intact skin.
- It is easier to handle canned coconut cream than the fresh one. It does not rot easily.

Ingredients

500 g taro
200 g sago
1 small can coconut cream (about 160 ml)
12 cups water

Seasoning

120 g sugar

One day before feast

It is not suitable to prepare the sweet soup ingredients in advance.

On the day of feast

1. Put 6 cups of water into a large pot. Bring to the boil. Add the sago and give a good stir. Turn off heat. Leave with a lid on for about 10 minutes. Stir the sago apart until they are transparent (if not fully transparent, turn on heat and bring to the boil. Then turn off heat and leave with a lid on until they are transparent). Put them in ice water. When they are cool, drain and set aside.
2. Peel and rinse the taro. Cut into small dices.
3. Bring 6 cups of water to the boil. Put in the taro and cook for about 15 minutes, or until tender. Season with the sugar. Stir until the sugar dissolves. Put in the sago. Stir and cook for a while. Finally pour in the coconut cream. Serve.

Sweet and Spicy Chicken

Tips for shopping

- Chilled chicken is a choice because this dish focuses more on the spiciness than the fresh flavour of the chicken. Better buy chilled chicken at about $60 to $70 each. It is more expensive but the quality is promising.
- Mini cucumbers have a thin body with tiny spines on the skin. They tend to be crunchier and sweeter as compared to thick and big cucumbers. They are suitable for making cold dishes.
- You can buy ready-made deep-fried peanuts to save work.

Ingredients

1/2 chicken
2 mini cucumbers
100 g deep-fried peanuts
1 tsp sesame seeds

Marinade

5 slices ginger
2 sprigs spring onion (sectioned)
3 tsps salt
2 star anises

Seasoning for cucumber

2 tsps chilli bean sauce
2 tsps sugar
2 tsps sesame oil
1 tsp finely chopped garlic

Sweet and spicy sauce

5 tbsps all-purpose marinade with herbs
2 tbsps chilli oil
2 tbsps Sichuan peppercorn oil
1 tbsp Hua Diao wine
1 tbsp Zhenjiang vinegar
3 tsps chilli bean sauce
1/2 tsp ground Sichuan peppercorns
1 tbsp sesame oil
1 tsp diced ginger
1 tbsp finely chopped garlic
1 tbsp diced spring onion
1 tbsp chopped coriander
1 tbsp diced red chilli

One day before feast

- Wash the chicken. Put the marinade into the chicken cavity. Rub evenly. Leave for about 1 hour. Steam over medium heat for half an hour, or until done. Take out. Let it cool down. Keep refrigerated.
- Mix the sweet and spicy sauce. Put into a glass jar. Keep refrigerated.

On the day of feast

1. Put the sesame seeds in a dry wok. Stir-fry over low heat until fragrant.
2. Rinse the mini cucumbers. Wipe dry. Cut lengthwise in half. Remove the seeds. Lightly pat with a knife. Cut into about 4 cm-long strips. Sprinkle with 2 tsps of salt and mix well. Leave for about 5 minutes. Put into ice water to remove the saltiness. Drain. Mix with the seasoning for cucumber. Rest for about 15 minutes. Put on a plate.
3. Take out the cooked chicken from the refrigerator. Rest at room temperature (or warm up in a microwave oven over low heat). Chop into pieces. Put on top of the cucumbers. Heat the sweet and spicy sauce. Pour on the chicken. Finally sprinkle with the stir-fried peanuts and sesame seeds. Serve.

Dried Scallop and
Winter Melon Thick Soup

Tips for shopping

- Select a ripe winter melon. It should be smooth to the touch with no cracks on the skin, or the skin has a layer of white powder.
- It is no need to use whole scallops. You can buy crumbled scallops, which are cheaper.
- Vacuum-packed Jinhua ham meat is cheaper and can be used.
- Chicken stock with scallop flavour is a choice. It can be in cans or boxes.

Ingredients

10 dried scallops (about 80 g)
1 kg winter melon
50 g Jinhua ham meat
3 cups chicken stock
1 cup water
2 egg whites

Seasoning

1 tsp salt
1/3 tsp ground white pepper

Caltrop starch solution

3 tbsps caltrop starch
6 tbsps water

One day before feast

185

- Skin the winter melon and rinse. Cut into small pieces. Dish up. Put in 2 slices of ginger. Steam for half an hour (no need to add water to the winter melon). Let it cool down. Pat into puree with a knife, or blend with a blender. Keep refrigerated.
- Rinse the Jinhua ham meat. Put into a bowl. Add water to cover the ham. Steam for about 40 minutes. Finely chop and keep refrigerated. Reserve the steamed ham sauce for cooking the soup.
- Soak the dried scallops in water (about 1 hour). Steam for about 45 minutes. Let it cool down. Tear into shreds. Keep refrigerated. Reserve the steamed scallop water for cooking.

On the day of feast

1. Put the chicken stock, steamed ham sauce, steamed scallop water and water together. Bring to the boil. Add the winter melon and scallops. Mix well. Bring to the boil. Put in the seasoning and slightly stir.
2. Mix the caltrop starch solution. Pour into the soup bit by bit while stirring the soup with a ladle. Cook until it is thick. When it comes to the boil, turn off heat. Add the egg whites and mix gently. Transfer to a big bowl. Finally sprinkle with the chopped ham. Serve.

Baked Mud Crabs in Shunde Style

Tips for shopping

- Having rich and sweet roe, female mud crab is most suitable for making this dish. Try to observe the abdomen of the crab. If it is protruding with a little roe, the crab is full of roe. Female mud crabs are most plump from June to August.
- Chicken egg can be replaced with duck egg, which has an intense egg flavour.
- As it is a baked dish, it is more suitable to use an earthen bowl. If the bowl is small, divide into two servings.

Ingredients

2 female mud crabs
(about 600 g each)
6 preserved olives
400 g pork collar-butt
80 g fat pork
1 piece dried tangerine peel
3 sprigs coriander
6 eggs

Spices

20 g diced ginger
1 tsp finely chopped garlic
50 g diced spring onion

Seasoning for stuffing

1 tsp salt
1 tsp chicken bouillon powder
2 tsps caltrop starch
1/2 tsp sesame oil
1/3 tsp ground white pepper

Seasoning

1/2 tsp salt
1 tbsp oyster sauce
4 tbsps water

One day before feast

- Soak the dried tangerine peel in water until soft. Scrape off the pith.
- Finely dice the tangerine peel, preserved olives and coriander stems. Keep refrigerated.
- Finely chop the pork collar-butt and fat pork. Keep refrigerated.

On the day of feast

1. Turn over the crab. Insert a chopstick into the abdomen until the crab is inactive. Untie the crab. Open the shell. Remove the internal organs and rinse. Reserve the roe.
2. Chop each side of the crab into 3 pieces. Cut the claw into 2 pieces (crack with a knife). Sprinkle with 3 tsps of caltrop starch.
3. Heat a wok. Put in 2 tbsps of oil. Stir-fry the spices until fragrant. Add the crabs and stir-fry over high heat for a while. Put in the seasoning. Give a good stir. Put a lid on. Adjust to medium heat. Leave for 2 minutes. Turn off heat. Transfer the crabs into an earthen bowl.
4. Preheat an oven to 200℃ for 10 minutes.
5. Whisk 6 eggs. Reserve the egg wash of 1 egg. Mix the rest egg wash with the seasoning for stuffing and the other ingredients (preserved olives, tangerine peel, corianders and pork). Finally add the crab roe and mix well.
6. Pour the egg mixture into the earthen bowl to cover the crabs. Steam for about 20 minutes. Take out (sop up any water with kitchen paper).
7. Spread egg wash on the surface of the dish. Bake in the oven for 10 minutes until golden. Take out. Spread with the rest egg wash. Bake for 5 minutes until the surface is crisp. Sprinkle with diced spring onion. Serve.

ried Shrimp Stuffed
hanghainese Bok Choy

ps for shopping

You may use fresh shrimps or frozen shelled shrimp. It requires much work for handling fresh shrimps. Frozen shelled shrimp, generally produced in Vietnam or Thailand, is available at supermarkets or frozen food shops in packets of 1 kg each. Do not choose live shrimps as it is difficult to remove the shells.

Select Shanghainese bok choy with a thick body for cooking.

Buy packed corn kernels. If they cannot be used up, keep them in a refrigerator. For corn kernels sold in cans, it is hard to store the leftovers. They may be wasted.

Ingredients

300 g shelled shrimps
10 stalks Shanghainese bok choy
50 g corn kernels
30 g diced carrot
1 egg white

Seasoning for minced shrimp

1/2 tsp salt
1/2 tsp chicken bouillon powder
3 tsps caltrop starch
1/2 tsp sesame oil
1/3 tsp ground white pepper
1/2 egg white

Seasoning for thickening glaze

1/2 cup chicken stock
1 tbsp caltrop starch
2 tbsps water

One day before feast

Method for minced shrimp:

1. Defrost the shelled shrimp (for fresh shrimps, remove the shell). Rub the shelled shrimp with caltrop starch. Rinse.
2. Sop up the water on the shelled shrimp with dry cloth. Pat the shelled shrimp with a knife. Chop with the back of the knife into puree.
3. Put the shrimp puree into a big bowl. Rub by hand for about 3 minutes, or until it is a bit sticky. Add the seasoning for minced shrimp. Stir in one direction (about 5 minutes) until it is sticky and spongy. Throw into the bowl repeatedly until it is gluey. Mix in the corn kernels and carrot. Keep refrigerated.

187

On the day of feast

1. Shave a few leaves off the Shanghainese bok choy. Cut in half and rinse. Blanch for about 2 minutes. Rinse in cold water.
2. Sop up the water on the Shanghainese bok choy. Spread a little caltrop starch on the surface. Fill in the stuffing to make it look like a small hill.
3. Heat a wok. Add 2 tbsps of oil. Put in the Shanghainese bok choy with the stuffed side down. Fry over low heat until done.
4. Pour the chicken stock into the wok. Simmer for a while (about 3 minutes). Arrange the Shanghainese bok choy on a plate. Thicken the chicken stock with caltrop starch solution. Bring to the boil. Sprinkle on the Shanghainese bok choy. Serve.

Stir-fried Shrimps in Two Flavours

Tips for shopping

- Pick green, thick and firm celery. The American type is crunchier and sweeter.
- For shrimps, buy either the fresh or frozen.

Ingredients

500 g shelled shrimps
400 g celery (strings removed; cut diagonally)
200 g spinach
100 g carrot (blended into juice)
100 g carrot (peeled; cut diagonally)

Spices (2 portions)

1 tsp finely chopped garlic
ginger slices

Seasoning for stir-frying

1 tsp salt
1 tsp chicken bouillon powder
1 tsp ginger juice
1 tbsp oil
1/2 cup boiled water

Seasoning for shrimps in spinach sauce

1/3 tsp salt
1/3 tsp chicken bouillon powder
1 tsp caltrop starch
1 tbsp spinach juice
1/3 tsp sesame oil

Seasoning for shrimps in ketchup

1/3 tsp salt
1/3 tsp chicken bouillon powder
1 tsp sugar
1 tsp caltrop starch
1 tbsp carrot juice
1 tbsp water
2 tbsps ketchup

One day before feast

Method for shrimps in spinach sauce:

1. Rub the shelled shrimps with caltrop starch and rinse. Sop up the water. Divide into two portions (250 g each). One is for cooking with spinach sauce. Another is for cooking in ketchup.
2. Rinse the spinach. Blanch for about half minute. Dish up. Cool in cold water. Drain.
3. Put the spinach into a blender. Add 3 tbsps of water and blend together. Sieve with a mesh strainer. Press juice out from the spinach residues.
4. Mix 4 tbsps of the spinach juice, 1/3 tsp of salt and 2 tsps of caltrop starch. Put in one portion of the shrimps and mix well. Keep refrigerated. Reserve the rest spinach juice.

Method for shrimps in ketchup:

1. Rinse 100 g of carrot. Cut into pieces. Put into a blender. Add 3 tbsps of water. Blend together into puree. Sieve with a mesh strainer. Press out the juice from the carrot residues.
2. Take another portion of the shrimps. Add 3 tbsps of the carrot juice, 1/3 tsp of salt, 2 tsps of caltrop starch and 1 tsp of sesame oil. Mix well. Keep refrigerated.

On the day of feast

1. Cut and rinse the celery and carrot. Heat a wok. Put in the seasoning for stir-frying. Add the celery and carrot and stir-fry for about 2 minutes. Drain. Divide into two portions.
2. Blanch the shrimps in spinach sauce and the shrimps in ketchup separately until done.
3. Heat a wok. Put in 1 tsp of oil. Stir-fry one portion of the spices over low heat until fragrant. Add the shrimps in spinach sauce. Roughly stir-fry. Put in the seasoning for shrimps in spinach sauce. Turn to medium heat and swiftly stir-fry for a while. Add one portion of the celery and carrot. Give a good stir-fry. Put on a plate.
4. Wash the wok. Heat up and add 1 tsp of oil. Put in another portion of the spices. Stir-fry over low heat until fragrant. Add the shrimps in ketchup and roughly stir-fry. Put in the seasoning for shrimps in ketchup. Turn to medium heat and swiftly stir-fry for a while. Add another portion of the celery and carrot. Give a good stir-fry. Put on the plate. Serve.

Stewed Taro and Duck in Chu Hou Sauce

Tips for shopping

- Chilled duck can be used. Select the duck with intact, yellowish skin and scarlet meat.
- Buy large-sized pickled plums at groceries. Some are packed in small bags (about 3 to 4 pieces per bag).

Ingredients

1 duck (about 1.5 kg to 1.8 kg)
1.5 kg taro
6 pickled plums (cored)
2 tbsps oil
5 cups water

Spices

60 g ginger slices
40 g diced shallots
1 tbsp finely chopped garlic
3 sprigs spring onion (sectioned)
3 star anises
1 piece dried tangerine peel (soaked to soft; pith scraped)

Seasoning

3 tbsps Chu Hou sauce
1 piece fermented tarocurd
2 tsps salt
1 tsp chicken bouillon powder
3 tbsps oyster sauce
3 tbsps dark soy sauce
3 tbsps crushed rock sugar
2 tbsps Shaoxing wine

One day before feast

Method for duck in Chu Hou sauce:

1. Gut and rinse the duck. Cut away the tail of the duck (remove the unpleasant smell). Cut off the legs and tip of the wings (reserve for stewing with the duck). Colour the duck by spreading 2 tbsps of dark soy sauce on the surface evenly.
2. Heat a wok. Put in 1 tsp of oil. Fry the duck over low heat until golden and fragrant.
3. Heat the wok. Put in 2 tbsps of oil. Stir-fry the spices over low heat until aromatic. Add the Chu Hou sauce and fermented tarocurd. Give a good stir-fry. Sprinkle with the Shaoxing wine. Roughly stir-fry. Add 5 cups of water, pickled plums and the rest seasoning. Cook until they dissolve. Put in the duck and bring to the boil. Turn to low heat and stew for about 1 hour (alternately turn the duck over to heat both sides evenly while stewing). Let it cool down. Put the duck with sauce in a refrigerator.

On the day of feast

1. Peel and rinse the taro. Cut into thick slices (about 1/2 inch thick). Fry in oil until golden. Set aside.
2. Put the duck with sauce into the wok. Heat up. Arrange the taro in the wok. The sauce should cover the taro (if there is not enough sauce, add boiled water). Simmer for about 40 minutes. Turn off heat. Leave with a lid on for about 10 minutes. Arrange the taro on a plate. Chop the duck into pieces. Put on top of the taro. Thicken the sauce with caltrop starch solution. Pour the sauce on top. Serve.

189

Braised E-Fu Noodles with Assorted Mushrooms and Shrimp Roe

Tips for shopping

- E-fu noodles are available at shops selling raw noodles. It is good to buy large patty-like E-fu noodles that are thick, round and golden without odd, greasy smell.
- Dried shrimp roe is sold in dried seafood groceries in bulks or small jars.

Ingredients

2 pieces dried E-fu noodles
2 bundles Enokitake mushrooms
60 g fresh black mushrooms
60 g oyster mushrooms
150 g mung bean sprouts
3 tsps dried shrimp roe
80 g yellow chives
2 tsps oil

Seasoning

1 tsp salt
1 tbsp oyster sauce
1/2 tsp chicken bouillon powder
1 tbsp dark soy sauce
1 1/2 cups water
1 tsp sesame oil

One day before feast

- Regular dried shrimp roe is raw. The shrimp roe bought need to be stir-fried in a dry wok until aromatic. Store the cool shrimp roe for cooking.
- Blanch the E-fu noodles to soften. Dish up. Let it cool down. Keep refrigerated.

On the day of feast

1. Cut the black mushrooms and oyster mushrooms into coarse strips. Section the Enokitake mushrooms and yellow chives. Rinse well.
2. Heat a wok. Put in 2 tsps of oil. Add 2 tsps of the shrimp roe and the assorted mushrooms. Stir-fry over high heat until fragrant. Put in the seasoning. Bring to the boil. Add the E-fu noodles and simmer for about 3 minutes.
3. Put in the mung bean sprouts and mix well. When the noodles are tender, add the yellow chives. Mix well. Put on a plate. Finally sprinkle with 1 tsp of the shrimp roe. Serve.

Ginger Milk Pudding

Tips for shopping

- Kowloon Dairy hi-calcium low fat milk is recommended. It is ideal to use fresh buffalo milk.
- Select thick and big ginger root. It is smooth to the touch with a crunchy texture. When the ginger breaks, there is juice oozing.

Ingredients

1 box hi-calcium low fat milk (large, about 1 kg)
400 g ginger

Seasoning

6 tsps sugar

191

One day before feast

It is not necessary to prepare this dessert in advance. It should be done right before serving.

On the day of feast

1. Peel and rinse the ginger. Grate with a ginger grater (if a blender is used, slice the ginger before blending).
2. Squeeze the juice from the grated ginger. Prepare 8 small bowls. Put 2 tbsps of the ginger juice into each bowl. Reserve the grated ginger for cooking.
3. Bring the fresh milk to the boil at 100℃ (keep stirring while cooking to avoid sticking on the pot). Add sugar and cook until it dissolves. Turn off heat. When the temperature of the milk reduces to around 75℃ to 80℃ , pour quickly into the bowls (80% full is enough). Wait for 5 minutes to set. Serve.

* If there is no food thermometer at home, I suggest this method to reduce the temperature: When the milk comes to the boil, pour it into another small pot or measuring cup, and then back to the former pot. Repeat this step for 4 times. Then quickly pour the milk into the bowls. Wait for 5 minutes to set.

Wasabi Shrimp Salad

Tips for shopping

- Either fresh or frozen shelled shrimps can be used.
- Better buy smooth, thick cucumbers because they are pulpy and crunchy. It is easier to peel into shreds.
- You can pick Kraft Miracle Whip for salad dressing.

Ingredients

20 shelled shrimps
2 cucumbers
3 eggs

Salad sauce

200 g salad dressing
2 tsps wasabi
1 tbsp condensed milk
2 tsps lemon juice

One day before feast

- Defrost the shelled shrimps (for fresh shrimps, remove the shell). Rub the shrimps with caltrop starch. Rinse well. Marinade with 1 tsp of salt for half an hour. Blanch until done. Cool in ice boiled water. Butterfly the shrimps. Remove the veins. Rinse with cold boiled water. Sop up the water. Keep refrigerated.
- Peel the cucumbers. Rub and wash with water. Wipe dry. Shred with a grater. Soak in ice water. Put in a dozen ice cubes. Keep refrigerated for about 3 hours. Drain and wipe dry. Keep refrigerated.
- Put the eggs with shell on into cold water. Boil until done (about 12 minutes). Cool in cold water. Remove the shell. Keep refrigerated.

On the day of feast

1. Mix the salad sauce. Keep refrigerated.
2. Separate the cooked egg yolks from the egg whites. Finely dice the egg whites. Put the egg yolks in a mesh strainer. Press out the powdery egg yolk with a small spoon.
3. Mix 3 tbsps of the salad sauce, shredded cucumber and diced egg whites. Put on a plate. Top with the shrimps. Put the rest salad sauce on top. Finally sprinkle with the powdery egg yolk. Serve.

Double-steamed Black-skinned Chicken with American Ginseng

Tips for shopping
- You can buy chilled black-skinned chicken.
- Pick full lean pork to avoid oil floating on the soup.
- Use only the meat of Jinhua ham. Vacuum-packed products can be used.
- Select American ginseng in whole sticks in herbal medicine shops, and let the shop assistant slice for you. The quality is more promising.

Ingredients

1 black-skinned chicken
300 g lean pork
40 g Jinhua ham meat
30 g sliced American ginseng
1 cup chicken stock
3 slices ginger
5 cups water

Seasoning

1 tsp salt
2 tsps Shaoxing wine

One day before feast
- Wash the black-skinned chicken. Chop into pieces. Blanch until done. Cool in cold water. Keep refrigerated.
- Dice the lean pork and Jinhua ham. Blanch until done. Cool in cold water. Keep refrigerated.
- Soak the sliced American ginseng in 1 bowl of hot water. Put a lid on. When it is cool, store the bowl of American ginseng with water in the refrigerator.

On the day of feast
1. Pour the water and chicken stock into a ceramic pot for double steaming. Put in the chicken, Jinhua ham, lean pork and ginger. Add the Shaoxing wine. Mix well.
2. Bring 5 cups of water in a wok to the boil. Put in the pot with a lid on. Double-steam for 1 1/2 hours (if there is insufficient water in the wok, add hot water).
3. Put in the American ginseng water. Mix well. Double-steam again for 1 hour. Season with salt. Serve.

Steamed Blue Crabs with Pickled Plums

Tips for shopping

Select blue crabs with active legs and eyes. The legs should be intact with no odd smell. If the legs and claws look transparent, it means the meat has not yet matured and is watery. If they appear cream in colour, the crab is meaty and good.

Ingredients

2 blue crabs (about 600 g each)
2 sprigs spring onion
2 sprigs coriander

Seasoning

6 pickled plums
2 tbsps plum sauce
2 tbsps ketchup
1/2 tsp salt
1 tbsp sugar
2 tsps finely chopped garlic
2 tsps diced ginger
2 tsps diced red chillies
2 tsps caltrop starch
2 tsps oil
4 tbsps water

One day before feast

Blue crab is a kind of seafood which should be bought on the day of cooking. The seasoning for steamed crabs can be prepared beforehand.

1. Core the pickled plums. Chop up.
2. Heat a wok. Add 2 tsps of oil. Stir-fry the garlic and ginger over low heat until aromatic. Put in all the seasoning and mix well. When it cools, keep refrigerated.

On the day of feast

1. Slightly rinse the crabs. Put the abdomen side up. Insert a sharp bamboo chopstick from the tip of the cover into the abdomen until the crab is inactive. Untie the crab. Open the shell. Remove the internal organs including the gills, stomach, and intestine. Wash thoroughly.
2. Chop each side of the crab body (with legs) into 3 pieces. Chop the claw into 2 pieces. Crack the claw with a knife so that it can take in flavour and be cooked through easily.
3. Arrange the crabs on a plate. Spread the seasoning on the crabs. Steam over boiling water for about 12 minutes. Take out. Sprinkle with the spring onion and coriander. Serve.

Baked Stuffed Mussels in French Style

Tips for shopping
- Buy frozen half-shelled mussels. You may choose the products from Australia or New Zealand. Their quality is more promising. They are generally sold in boxes of about 25 pieces each at supermarkets or frozen food shops.
- It is more convenient to use packed cheese slices.

Ingredients

20 frozen mussels
200 g minced pork
100 g button mushrooms
100 g celery
100 g onion
100 g carrot
4 slices cheese
1 egg

Cream sauce

60 g butter
40 g flour
4 tbsps water
3 tbsps fresh milk

Seasoning

1 tsp salt
1 tsp chicken bouillon powder
1 tsp sugar
1 tsp mixed herbs
1 tbsp curry paste

One day before feast
Method for mussel stuffing:
1. Remove the shell from mussel. Rub the shell clean and reserve the shell. Rinse and coarsely dice the meat. Blanch until done.
2. Finely dice the button mushrooms, celery, carrot and onion. Blanch until done.
3. Mix the minced pork with 1/3 tsp of salt. Stir-fry in a wok over low heat until done.
4. Put the butter in the wok. Melt over low heat. Add the flour and stir-fry until fragrant. Put in the water and fresh milk. Mix well. The cream sauce is done.
5. Heat the wok. Add 1 tsp of oil. Stir-fry the onion until aromatic. Put in the cheese and cook until it melts. Add the other ingredients (except egg). Give a good stir-fry. Put in the seasoning and cook until it dissolves. Finally add the cream sauce and mix well. The stuffing is done. Let it cool down. Keep refrigerated.

195

On the day of feast
1. Preheat an oven for 10 minutes, with upper heat about 200℃ and lower heat 180℃ .
2. Blanch the mussel shells to make hot. Drain and wipe dry.
3. Fill the stuffing into the shell in full. Brush the whisked egg on the surface. Bake for about 7 minutes. Take out. Brush with the egg wash again. Bake for 3 minutes. Put on a plate and serve.

Dual-Flavours Scallops with Vegetables

Ingredients

600 g scallops

Marinade

1/3 tsp chicken bouillon powder
2 tsps caltrop starch
1 tsp sesame oil
1/4 tsp ground white pepper

Deep-fried scallops
Ingredients:

200 g bread crumbs

Egg batter

1 egg
4 tbsps caltrop starch
* mixed well

Stir-fried scallops
Ingredients:

600 g broccoli
150 g fresh straw mushrooms

Spices

finely chopped garlic
ginger slices

Seasoning for scallops

1/3 tsp salt
1/3 tsp chicken bouillon powder
1 tsp caltrop starch
1 tbsp water

Seasoning for broccoli

1 tsp salt
1/2 tsp chicken bouillon powder
1 tsp oyster sauce
1 tsp caltrop starch
1 tbsp boiled water

Tips for shopping

- You can use frozen scallops. Quality products from Australia are more expensive. Bigger in size, Japanese scallops are cheaper but the quality is not so good.
- Better use Japanese coarse bread crumbs for deep-frying. It is crunchier with a golden tint.

One day before feast

Defrost the scallops at room temperature. Remove the substance on the edge of the scallops. Gently rinse and drain. Mix with the marinade. Divide into two portions. Keep refrigerated.

On the day of feast

1. Cut the broccoli into small floral. Cut the straw mushrooms in half. Rinse well.
2. Take one portion of the scallops. Slightly blanch in boiling water. Dish up. Cool in cold water. Sop up the water. Dip in the egg batter and then coat with the bread crumbs.
3. Take another portion of the scallops. Gently blanch. Turn off heat. Leave for about 1 minute to let the heat of hot water make it done. Drain.
4. Put the broccoli and straw mushrooms into boiling water. Add 1 tsp of salt and 2 tsps of oil. Blanch until done and drain. Put into a wok. Add the seasoning for broccoli. Swiftly stir-fry over medium heat. Put the broccoli on the side of a plate. Arrange the straw mushrooms in the middle.
5. Heat the wok. Put in 1 tsp of oil. Stir-fry the spices until fragrant. Add the scallops from Step 3. Give a good stir-fry. Sprinkle with the seasoning for stir-fried scallops. Quickly stir-fry over medium heat. Sprinkle with Shaoxing wine. Put on top of the straw mushrooms.
6. Heat oil in the wok to medium hot (about 180 ℃ to 200 ℃). Deep-fry the scallops coated with bread crumbs until done and golden. Put on the plate. Serve.

197

Dried Oyster and Preserved Meat Wrapped in Lettuce

Tips for shopping

- Select dried oysters which are dry, intact, and plump. The products from Japan or South Korea are better.
- Preserved pork sausages and liver sausage with an even distribution of fat and lean meat can be used. The good type is dry to the touch with no oil oozing from the skin. If the fat pork on the surface layer is yellowish, the sausage is not fresh.
- Buy corn flakes in original flavour, which is light, at supermarkets.

Ingredients

15 dried oysters
2 preserved pork sausages
1 preserved liver sausage
150 g minced pork collar-butt
150 g Chinese celery
10 dried black mushrooms
10 skinned water chestnuts
2 sprigs spring onion (diced)
120 g corn flakes
2 iceberg lettuces

Seasoning for stewing

2 cups water
1 tsp salt
1 tbsp oyster sauce
1 tsp sugar
1 tsp cooking wine (Hua Diao or double distilled rice wine)
3 slices ginger
1 tbsp oil

Spices

1 tsp diced ginger
1 tsp finely chopped garlic

Seasoning for minced pork

2 tsps oyster sauce
1 tsp caltrop starch
2 tsps water

Seasoning for filling

1/2 tsp salt
1 tsp chicken bouillon powder
1 tbsp oyster sauce
1 tbsp dark soy sauce
1 tsp caltrop starch
1 tsp sesame oil
1/3 tsp ground white pepper
2 tbsps boiled water

One day before feast

Method for dried black mushrooms and dried oysters:

1. Soak the dried black mushrooms and oysters in water for about 3 hours to soften. Rinse well.
2. Bring the seasoning for stewing to the boil. Put in the black mushrooms. Simmer for about 1 hour. Add the dried oysters and cook together for about 10 minutes. Dish up. Finely dice when cool. Keep refrigerated.

Method for preserved pork sausages and liver sausage:

Slightly blanch the preserved meat to remove grease and dirt. Steam the meat for about 8 minutes. Take out. Finely dice when cool. Keep refrigerated.

On the day of feast

1. Tear the leaves off the iceberg lettuces. Soak in water to clean. Trim into a shallow dish-like shape. Wipe dry.
2. Rinse and finely dice the Chinese celery. Finely chop the water chestnuts. Squeeze out water.
3. Mix the minced pork with the seasoning. Heat a wok. Put in 2 tsps of oil. Stir-fry the minced pork over medium heat until done. Set aside.
4. Heat the wok. Put in 2 tsps of oil. Add the spices and preserved meat. Stir-fry over low heat until it is fragrant and oil oozes. Add the oysters and minced pork. Stir-fry until fragrant. Turn to medium heat. Put in the Chinese celery, black mushrooms and water chestnuts. Swiftly stir-fry until hot.
5. Mix the seasoning for filling. Gradually put into the wok. Stir-fry evenly. Finally sprinkle with the Shaoxing wine. Set aside.
6. Lay the corn flakes on a plate. Top with the filling. Sprinkle with the spring onion. Wrap in the lettuce to serve, or spread with a little Hoi Sin sauce.

Stir-fried Rice with Seafood and Pineapple

Tips for shopping

- A good pineapple has golden skin with big eyes (scraggly dots). It smells aromatic. If the leaves have fallen from the top and the pineapple smells fermented like wine, it is ripe and likely to rot.
- Both the shelled shrimps and scallops can be frozen products. A fresh squid has protruding eyes and a glossy, slightly transparent body. The skin is also intact.

Ingredients

1 fresh pineapple (large)
10 shelled shrimps
1 fresh squid
10 frozen scallops
5 imitated crab sticks
40 g carrot
8 strips flowering Chinese cabbage
1 sprig spring onion
2 eggs
1 tbsp oil
4 bowls steamed rice

Marinade

1/3 tsp salt
2 tsps caltrop starch
1 tsp sesame oil
1/4 tsp ground white pepper

Seasoning

1 tsp salt
1 tsp chicken bouillon powder
2 tsps light soy sauce

One day before feast

- Shave the leaves off the pineapple. Cut lengthwise in half. Extract the flesh with a small knife or pulp scraper (be careful not to break the pineapple shell). Soak the flesh and shell separately in cold salted water for about 10 minutes. Wipe dry. Seal with cling wrap and keep refrigerated. It is to prevent the flesh from changing colour as a result of oxidation.
- Defrost and dice the shelled shrimps and scallops. Cut open the squid belly. Remove the internal organs and skin. Rinse and cut into dices. Wipe the seafood dry. Mix with the marinade. Keep refrigerated.

On the day of feast

1. Dice the crab sticks and pineapple flesh. Finely slice the carrot and flowering Chinese cabbage. Rinse well. Dice the spring onion.
2. Take out the pineapple shell from the refrigerator. Soak in warm water to warm up. Wipe dry.
3. Blanch the seafood until done.
4. Heat a wok. Put in 1 tbsp of oil. Stir-fry the flowering Chinese cabbage and carrot until fragrant. Add the whisked eggs and slightly stir-fry. Pour in the steamed rice. Give a good stir-fry.
5. Add the seafood and pineapple flesh. Swiftly stir-fry over medium heat until hot through. Sprinkle with the seasoning. Quickly stir-fry over high heat until even. Finally sprinkle with the spring onion. Slightly stir-fry. Transfer into the pineapple shell. Serve.

199

Almond Sweet Soup with White Fungus and Egg Whites

Tips for shopping

- With a light almond flavour, sweet almond is bigger and plumper than bitter almond. As the latter has a bitter taste, complement the sweet almonds with a little amount of bitter almonds to enhance the aroma.
- The white fungus from Zhangzhou is better in quality. It is big in size and yellowish. Its frond-like body scrolls closely like a ball. Some of the cultivated white fungi offered in the market are even bigger. They are yellowish with a tint of white, and the body scrolls loosely. After rehydration, they turn soft and will be cooked all to a mash easily.

Ingredients

300 g sweet almonds
40 g bitter almonds
40 g white fungus
6 cups water
8 egg whites
1/2 cup fresh milk
3 slices ginger

Seasoning

250 g rock sugar

One day before feast

- Rinse the almonds. Soak in 2 cups of water for about 2 hours. Blend the almonds and water together with a blender. Strain with a mesh strainer. Press out the juice from the almond residues. Keep the almond juice refrigerated.
- Soak the white fungus in water for about 2 hours to soften. Cut away the hard, yellow stalk. Rinse well. Put in boiling water with ginger juice. Boil for 10 minutes. Cool in cold water. Cut into small pieces. Keep refrigerated.

On the day of feast

1. Bring 4 cups of water to the boil. Add the white fungus and ginger. Turn to low heat and cook for about 40 minutes. Pour in the almond juice and cook for about 10 minutes. Season with the rock sugar until it dissolves. Turn off heat.
2. Whisk the egg whites. Slowly pour into the sweet soup. Stir evenly. Finally mix in the fresh milk. Serve.

Celery with Shredded Chicken

Tips for shopping

You can also use the frozen chicken steak. The boned chicken leg is meaty and more convenient for cooking.

Ingredients

3 chicken steaks
300 g celery
100 g carrot
4 sprigs coriander
1 deep-fried dough stick
1 red chilli
3 slices Chinese lettuce
1 tsp white sesame seeds (toasted)

Marinade

1/2 tsp salt
1/3 tsp chicken bouillon powder
3 tsps caltrop starch
3 tbsps water

Cold sauce

1 tbsp chilli bean sauce
1/2 tsp salt
1 tbsp sugar
1 tsp finely chopped garlic
1 tbsp sesame oil
2 tsps Zhenjiang vinegar
1 tbsp boiled water

One day before feast

- Remove the skin and fat of the chicken steak. Rinse well. Mix with the marinade and rest for about 1 hour. Cover the chicken steak with boiling water. Turn on low heat. Cook with a lid on for about 20 minutes, or until done. Cool in cold boiled water. Wipe dry. Tear into shreds. Keep refrigerated.
- Finely shred the celery and carrot. Rinse well. Blanch for 1 minute. Cool in ice water and drain. Keep refrigerated.
- Divide the deep-fried dough stick into two. Finely slice. Set aside.

201

On the day of feast

1. Rinse the coriander. Cut into small sections. Shred the red chilli.
2. Rinse the Chinese lettuce. Lay on a plate.
3. Deep-fry the dough stick in oil over medium heat until crisp. Dish up. Sop up the oil.
4. Mix the cold sauce. Put into a big bowl. Add the celery, carrot and shredded chicken. Mix well. Put in the sliced dough stick, coriander and red chilli. Mix well. Put on the plate. Finally sprinkle the white sesame seeds on top. Serve.

Double-steamed Chicken Soup with Abalones, Dried Scallops and Fish Maw

Tips for shopping

- Select thin-bodied fish maws (about 20 to 24 pieces per 600 g), which are cheaper. It is also easy for us to handle during the rehydrating process. Pick golden fish maws with round tubes and little cracks in the body.
- Frozen sea cucumbers, most of them rehydrated and offered in reasonable price, are the choice. It is easy for us to process. Better select full-bodied sea cucumber that is plump and meaty. It is more suitable to buy rehydrated sea cucumber weighting about 600 g to 900 g each.
- Canned baby abalones in soup can be used. It is better to choose one containing about 10 to 12 abalones. It is not chewy if the abalone is too small.
- Both the black-skinned chicken and pork shin have less fat. The soup will not be too greasy.
- Pick S-grade dried scallops (small ones, about 100 to 120 pieces per 600 g). Quality scallops are golden, intact, and aromatic.

Ingredients

10 dried whole scallops
3 fish maws
1 frozen sea cucumber
10 dried black mushrooms
1 can baby abalones in soup
80 g Jinhua ham
300 g pork shin
1 black-skinned chicken
400 g Peking cabbage
2 slices ginger
5 cups water
2 cups chicken stock

Seasoning

1 tsp salt
1 tsp Hua Diao wine

One day before feast

- Cover the dried scallops with water. Soak until they swell. Store the scallops with water together. Keep refrigerated.
- Soak the fish maws in water to soften (about 3 hours). Drain. Pour boiling water into a large pot. Soak the fish maws with a lid on until the water cools. Put into cold water to wash away the impurities on the surface of the fish maws. Boil in water with ginger juice to remove the odd smell. Cool in cold water and drain. Cut into pieces. Keep refrigerated.
- Defrost the sea cucumber. Boil in water with ginger juice wine. Cool in cold water. Cut into small pieces. Keep refrigerated.
- Soak the dried black mushrooms in water for about 3 hours. Remove the stalks. Rub with caltrop starch. Rinse well. Keep refrigerated.

On the day of feast

1. Cut away the top leaves of the Peking cabbage. Cut into quarters. Rinse and scald in boiling water.
2. Wash the chicken. Chop into pieces. Rinse the pork shin and Jinhua ham. Cut into dices. Scald in boiling water and rinse.
3. Put the chicken, Jinhua ham, pork shin and ginger into a ceramic pot for double steaming. Lay orderly. Surround by the Peking cabbage. Lay the black mushrooms, sea cucumber, fish maws and baby abalones on top.
4. Bring the chicken stock, dried scallop soaking liquid and 5 cups of water to the boil. Add 1 tsp of Shaoxing wine. Mix well. Pour into the ceramic pot. Arrange the whole scallops on top. Put a lid on. Double-steam over low heat for about 4 hours. Season with salt. Serve.

* It is most suitable to use a big ceramic pot for double-steaming this soup. If there is no ceramic pot at home, cook the soup in a regular soup pot over low heat, but the soup will not look as clear as the double-steamed.

Tips for shopping

- Look for live Mandarin fish. Chilled product is not recommended.
- Use regular dried tangerine peel. It is not necessary to buy expensive one.
- Pick seafood soy sauce. It will not be too salty.

Steamed Mandarin Fish with Mushroom and Jinhua Ham

Ingredients

1 Mandarin fish (about 600 g)
50 g lean pork
3 dried black mushrooms
20 g Jinhua ham meat
1 piece dried tangerine peel
10 g shredded ginger
1 sprig spring onion
2 sprigs coriander

Marinade

1/4 tsp salt
1 tsp oyster sauce
1 tsp caltrop starch
1 tsp sesame oil
2 tsps water
1/3 tsp ground white pepper

Seasoning

4 tbsps light soy sauce
2 tsps dark soy sauce
4 tbsps water
1/2 tsp chicken bouillon powder
3 tsps sugar
1/2 tsp sesame oil
1/4 tsp ground white pepper
2 sprigs coriander (pick stems)

One day before feast

- Shred the lean pork. Mix with the marinade. Keep refrigerated.
- Put the Jinhua ham meat into boiling water. Cook over low heat for 25 minutes. Drain. Let it cool down. Cut into shreds.
- Soak the dried black mushrooms in water to soften. Remove the stalks. Rub with caltrop starch and rinse. Cook them for about 25 minutes. Let it cool down. Shred the black mushrooms. Keep refrigerated.
- Soak the dried tangerine peel in water to soften. Scrape off the pith. Finely shred.

On the day of feast

1. Gill and rinse the fish. Wipe dry. Cut a slit on the meaty part (dorsal surface) of the fish.
2. Rinse the spring onion. Cut into shreds. Rinse the coriander. Reserve the leaves. Take the stems for cooking with the seasoning.
3. Cook the seasoning until it is hot and the sugar dissolves. Set aside.
4. Combine the pork, Jinhua ham, black mushrooms, tangerine peel and ginger together. Lay evenly on the fish.
5. Pour 4 cups of boiling water into a wok. Put in the fish. Cover with a lid. Steam over high heat for about 10 to 12 minutes. Take out. Remove excess liquid. Sprinkle with the spring onion and coriander. Pour 1 tbsp of boiling oil and the seasoning on top. Serve.

203

Stir-fried Shrimps with Egg Whites

Tips for shopping

- Either fresh shrimps or frozen shrimp meat can be used. It requires much work to process the fresh shrimps. Frozen shrimp meat is available in packets of 1 kg each.
- Only a little Jinhua ham meat will be used for this dish. Vacuum-packed product can be used.
- You may use cream instead of fresh milk to suit your taste. It has a rich creamy flavour.

Ingredients

600 g broccoli
400 g shelled shrimps
10 egg whites
150 ml fresh milk
20 g Jinhua ham meat

Marinade

1/2 tsp salt
1/2 tsp chicken bouillon powder
3 tsps caltrop starch
1 tsp sesame oil
1/4 tsp ground white pepper

Seasoning for egg whites

1/2 tsp salt
1/3 tsp chicken bouillon powder
ground white pepper
3 tsps caltrop starch
4 tsps water

Seasoning for broccoli

1 tsp salt
1/2 tsp chicken bouillon powder
1 tsp caltrop starch
2 tsps water

One day before feast

- Defrost the shelled shrimps. Rub with caltrop starch and rinse. Sop up the water. Mix with the marinade. Keep refrigerated.
- Put the Jinhua ham in boiling water. Cook over low heat for about 25 minutes. Let it cool down. Finely chop. Set aside.

On the day of feast

1. Cut the broccoli into small floral. Rinse well.
2. For the seasoning for egg whites, first mix the caltrop starch and water until the former dissolves. Add the rest seasoning. Mix well. Pour into the fresh milk. Give a good stir. Gently stir in the egg whites.
3. Heat 2 cups of water in a wok. Put in the broccoli and 1 tsp of salt. Blanch until it is 80% soft. Dish up. Stir-fry with the seasoning for broccoli until fragrant. Lay the broccoli on a plate.
4. Scald the shrimps in boiling water. Slightly deep-fry in oil over medium heat. Set aside.
5. Mix the milk and egg white mixture evenly. Pour into the wok with the cooked oil. Stir-fry over low heat until the egg white is 80% done. Put in the shrimps. Stir-fry together until fully cooked. Put on the plate. Sprinkle with the Jinhua ham. Serve.

Braised Pork Knuckle with Chinese Lettuce

Tips for shopping

Pork knuckle is the knee of a pig. It is sold at meat stalls, generally reserved. Specify that you want a boned pork knuckle. The shop assistant may burn down the tiny hairs on the skin for you. You just need to wash it at home.

Ingredients

1 pork knuckle
(about 1.2 kg to 1.5 kg)
600 g Chinese lettuce

Seasoning

2 tsps salt
1 tsp chicken bouillon powder
3 tbsps oyster sauce
20 g rock sugar
1 tbsp dark soy sauce
1 tbsp Hua Diao wine
4 star anises
1 piece dried tangerine peel
3 slices ginger
2 sprigs spring onion
4 cloves skinned garlic
4 cups water

One day before feast

Method for pork knuckle:

1. Wash the pork knuckle. Cook in boiling water for about 15 minutes. Rinse and wipe dry. Spread 2 tbsps of dark soy sauce on the skin evenly for colouring. Put 3 tbsps of oil into a wok. Fry the pork knuckle over medium heat until the skin is golden.
2. Put the ginger, spring onion and garlic in the wok. Stir-fry over medium heat until fragrant. Add the rest seasoning and water. Bring to the boil. Pour into a big bowl. Put in the pork knuckle. Steam for about 2 1/2 hours, or until tender. Let it cool down. Keep refrigerated.

On the day of feast

1. Steam the pork knuckle until hot (about 20 minutes). Put into a deep dish. Reserve 1 cup of the braised sauce as thickening glaze.
2. Cut away the head of the Chinese lettuce. Rinse well. Blanch until done. Put on the side of the pork knuckle.
3. Bring the braised sauce to the boil. Mix 3 tsps of caltrop starch with boiled water. Put into the sauce bit by bit while stirring to make it thick. Pour on the pork knuckle. Serve.

205

Crisp Chicken with Sand Ginger

Tips for shopping

- The chicken used for this dish can be chilled. Among the chilled ones, Kamei chicken is more well-known and its quality is promising. Its meat texture and flavour are similar to that of a fresh chicken. When buying, pay attention to the packing of the chicken. The bag containing the chicken should be intact and locked tightly, and stored in a refrigerator at 0°C to 4°C.
- Raw sand ginger has a richer aroma as compared to ground sand ginger. The sand ginger root is small with skin on. Both ingredients can be bought at Southeast Asian food groceries. The ground sand ginger offered by some groceries in small packs is seasoned powder for salt-baked chicken. It tends to be salty. It is no need to apply the seasoning if such product is used.

Ingredients

1 chicken (about 1.5 kg)
3 sprigs spring onion
4 cloves skinned garlic
50 g raw sand ginger

Marinade

3 tbsps salt
1 tbsp chicken bouillon powder
1 tbsp sugar
3 tbsps ground sand ginger

Ingredients for colouring chicken skin

4 tbsps white vinegar
1 tbsp maltose syrup

Dipping sauce

2 tbsps ground sand ginger
1 tbsp sesame oil
2 tbsps hot oil
1/2 tsp salt
1/3 tsp chicken bouillon powder
* mixed well

One day before feast

Method for chicken with sand ginger:

1. Mix the ingredients for colouring the chicken skin. Melt by sitting on hot water.
2. Rest the chilled chicken at room temperature to warm up. Rinse and wipe dry.
3. Section the spring onion. Wash the raw sand ginger and coarsely dice. Pat with the garlic together. Mix all these spices. Put into the chicken cavity.
4. Mix the marinade. Spread evenly on the chicken inside and outside. Leave for about 1 hour.
5. Bring water to the boil. Steam the marinated chicken for about 25 minutes. Remove the spices from the chicken cavity. Let it cool for about half an hour.
6. Absorb the oil and water on the chicken skin with kitchen paper. Spread the white vinegar and maltose syrup mixture on the skin. Hang the chicken until the skin is dry (about 4 to 5 hours). Wrap the chicken in cling wrap. Keep refrigerated.

On the day of feast

1. Chop the chicken in half. Rest at room temperature for about 1 hour.
2. Put 5 cups of oil into a wok. Turn on medium heat. When the oil is hot, adjust to low-medium heat. Deep-fry half portion of the chicken until the surface is slightly golden (about 5 minutes). Turn to high heat. Deep-fry again for 5 minutes, or until golden. Drain the oil. Put in another half of the chicken. Deep-fry in the same way. Chop into pieces. Put on a plate. Serve with the dipping sauce.

Glutinous Rice with Preserved Meat

Tips for shopping

- Look for preserved meat which is dry to the touch and has a large proportion of lean meat to fat.
- Select golden, dry, intact, and fragrant dried shrimps.

Ingredients

400 g glutinous rice
2 preserved pork sausages
1 preserved liver sausage
1/2 strip preserved pork
8 dried black mushrooms
150 g dried shrimps
1 sprig coriander (diced)
1 sprig spring onion (diced)
1 egg

Seasoning

2 tbsps light soy sauce
2 tbsps dark soy sauce
1 tbsp oyster sauce
1 tsp chicken bouillon powder
3 tbsps boiled water
sesame oil
ground white pepper

One day before feast

Method for glutinous rice:

1. Wash the glutinous rice. Cover with water. Soak for about 3 hours. Drain.
2. Lay a clean gauze cloth on a bamboo steamer or stainless steel colander. Put the glutinous rice evenly on the cloth. Steam for about 30 minutes (no need to add water to the glutinous rice). Let it cool down. Cover tightly with cling wrap. Keep refrigerated.

207

Method for preserved meat, dried shrimp and mushroom:

1. Slightly blanch the preserved meat. Steam until done (about 10 minutes). Cut into dices. Keep refrigerated.
2. Soak the dried shrimps in water to soften. Finely dice.
3. Soak the dried black mushrooms in water until soft. Remove the stalks. Rub with caltrop starch and rinse. Put into a wok. Cover with water. Season with 1 tsp of salt. Boil for about 20 minutes. Cut into dices. Keep refrigerated.

On the day of feast

1. Whisk the egg. Heat a non-stick pan over low heat. Pour in the egg wash evenly. Fry over low heat into egg skin. Cut into shreds.
2. Pour warm water (about 45 ℃) into a big bowl. Put in the steamed glutinous rice. Loosen the glutinous rice by hand and wash away the stickiness partially. Drain in a colander.
3. Heat a wok. Put in 1 tsp of oil. Stir-fry the preserved meat over low heat until it is aromatic and oil oozes. Put in the dried shrimps and black mushrooms. Give a good stir-fry. Put in the glutinous rice. Quickly stir-fry over medium heat until hot. Mix the seasoning and add in bit by bit. Stir-fry evenly.
4. Turn to high heat. Swiftly stir-fry for a moment. Put on a plate. Sprinkle with the diced spring onion and coriander. Lay the shredded egg on top. Serve.

Pear Sweet Soup with White Fungus and Lotus Seeds

Tips for shopping

- Look for pears with stalks and yellow skin with a green tint. It should be firm to the touch. If the body looks brown, the pear is overripe or starts to rot.
- Buy cored red dates for easy handling.
- Buy skinned whole lotus seeds produced from Xiantan, Hunan. It is easy for us to handle. It also smells beautifully.
- It is better to use the white fungus from Zhangzhou. It looks smaller than the cultivated one with a yellow tint. With a crunchy texture, it will not smash up easily while cooking.

Ingredients

5 pears
3 ears white fungus
20 red dates
100 g lotus seeds
4 slices ginger
8 cups water

Seasoning

250 g rock sugar

One day before feast

- Soak the lotus seeds in water to soften (about 2 hours). Pick out the cores with a bamboo stick to avoid the bitter taste. Rinse and keep refrigerated.
- Cover the white fungus with water. Soak until it swells (about 3 hours). Cut away the hard stalks and rinse. Put into boiling water. Add 1 tbsp of ginger juice with wine. Cook thoroughly (about 5 minutes) to remove its odd smell. Dish up. Finely chop into pieces. Let it cool down. Keep refrigerated.

On the day of feast

1. Rinse the red dates and ginger.
2. Bring 8 cups of water in a pot to the boil. Put in the white fungus, red dates, lotus seeds and ginger. Bring to the boil. Turn to low heat. Simmer for about half an hour.
3. Peel and core the pears. Cut the flesh into coarse strips. Rinse well. Put into the pot and cook for 40 minutes. Season with rock sugar. Cook until the sugar dissolves. Serve.

Deep-fried Spiced Tofu

Tips for shopping

- The spices for making Chinese marinade can be bought at regular Chinese herbal medicine shops. The cost of about 80 g of the spices is $10.
- It is most suitable to buy cloth-wrapped tofu which has a silkier texture compared with firm tofu. It also looks beautiful.
- Buy the preserved Sichuan vegetable in whole. To save work, you may buy packed Sichuan vegetable shreds in supermarkets or groceries (about 80 g per pack).

Ingredients

3 pieces cloth-wrapped tofu
1 whole (or 1 pack)
preserved Sichuan vegetable

Chinese marinade

40 g spices for Chinese marinade
1 tbsp salt
2 cups water

Dipping sauce

3 tbsps white vinegar
3 tbsps warm water
1 tbsp sugar
2 cloves skinned garlic

One day before feast

Method for spiced tofu:

1. Steam the spices with 2 cups of water for about half an hour to let the aroma spread.
2. Put in 1 tbsp of salt and the tofu. Steam over medium heat for about 1 hour, or until the surface of tofu is a bit rough. Remove and drain.
3. When the tofu cools, gently press water out with hands. Dry with kitchen paper or cloth. Put into an airtight container. Chill overnight for deep-frying the next day.

Method for preserved Sichuan vegetable slices:

1. Wash away the marinade on the surface of the preserved Sichuan vegetable. Finely slice. Soak in cold water for about 10 minutes. Drain.
2. Mix the Sichuan vegetable slices with 2 tsps of sugar and 1 tsp of sesame oil. Keep in a refrigerator.
* No preparation is needed for the packed Sichuan vegetable shreds. Use it right at cooking.

On the day of feast

1. Prepare the dipping sauce first: Finely chop the garlic. Mix with the other ingredients.
2. Put 3 cups of oil in a wok over high heat. When it is hot, adjust to medium heat. Deep-fry the spiced tofu for about 3 minutes. Turn up the heat and deep-fry for about half minute, or until the surface is golden. Drain.
3. Lay the sliced preserved Sichuan vegetable on a plate. Cut each piece of the deep-fried tofu into about 6 slices. Put on top of the Sichuan vegetable. Serve with the dipping sauce.

White Fungus and Mushroom Thick Soup

Tips for shopping

- Pick white fungus in light yellow instead of white. White fungus from Zhangzhou is a good choice.
- Select Enokitake mushrooms with thicker stems and soft, big umbrellas.
- Buy whole water chestnuts with skin on. Vacuum packed skinned water chestnuts are not promising in quality.

Ingredients

2 ears white fungus
10 dried black mushrooms
120 g oyster mushrooms
2 bundles Enokitake mushrooms
100 g carrot
2 small packs mung bean vermicelli
5 cups vegetarian stock

Ingredients of vegetarian stock

500 g carrot
30 dried black mushrooms
300 g soy bean sprouts
20 skinned water chestnuts
300 g fresh straw mushrooms
4 big slices ginger
12 cups water

Seasoning

1 1/2 tsps salt
1 tsp vegetarian chicken bouillon powder
1 tsp dark soy sauce

Caltrop starch solution

2 tbsps caltrop starch
4 tbsps water
* mixed well

One day before feast

Method for vegetarian stock:

1. Soak the dried black mushrooms in water to soften (about 1 hour). Cut away the stalks. Rub the black mushrooms and stalks with caltrop starch. Rinse well. Set aside.
2. Skin and rinse the water chestnuts. Peel the carrot. Cut into slices.
3. Put the straw mushrooms in boiling water. Add 1 tbsp of ginger juice. Blanch for about 5 minutes. Rinse well. Set aside.
4. Bring 12 cups of water to the boil. Put in all the ingredients. Turn to low heat and simmer for about 1 1/2 hours. Take out the black mushrooms, straw mushrooms and water chestnuts (as food ingredients of the vegetarian feast). When the vegetarian stock cools, keep in a refrigerator.

* Method for ginger juice:

Blend 300 g of skinned ginger with 300 ml of water. Remove and squeeze liquid out as ginger juice. Put in an airtight bottle and keep refrigerated. It can be kept for about 1 month.

On the day of feast

1. Soak the white fungus in water until it softens and swells up (about 1 hour). Cut away the yellow hard stalks and rinse. Put into boiling water with 1 tbsp of ginger juice. Boil for about 30 minutes. Rinse well. Cut into flakes. Set aside.
2. Cut the black mushrooms (taken from the vegetarian stock), oyster mushrooms and carrot into fine shreds. Set aside.
3. Soak the mung bean vermicelli in water to soften. Cut into short sections. Drain.
4. Cut away the end of the Enokitake mushrooms. Loosen and rinse.
5. Bring the vegetarian stock to the boil. Put in all the ingredients. Turn to medium heat and cook for about 5 minutes. Add the seasoning. Swiftly stir with a ladle. Slowly mix in the caltrop starch solution. Bring to the boil. Serve.

Hairy Melon Rings with Egg Tofu

Tips for shopping

- Select hairy melons with a straight and thinner body. The best thickness is around 2 inches to 2.5 inches in diameter.
- Pick green broccoli with compact flowers.

Ingredients

4 hairy melons
4 strips egg tofu
600 g broccoli

Seasoning for hairy melon rings

1 tsp salt
1 tsp sugar
1 tsp oil
1 tsp cooking wine
 (rice wine or Shaoxing wine)

Seasoning for broccoli

1 tsp salt
1 tsp sugar
1 tsp oil
1 tsp cooking wine

Seasoning for thickening glaze

1/3 tsp salt
1/2 tsp vegetarian chicken bouillon powder
1 tsp vegetarian oyster sauce
1/3 tsp dark soy sauce

One day before feast

Method for hairy melon rings:

1. Skin and rinse the hairy melon. Cut into round pieces of about 2 cm thick, looking like a chessman. Run a round mold or small knife in the middle of the hairy melon ring. Hollow out the ring (about 2.5 cm in diameter).
2. Put 3 cups of water in a wok. Bring to the boil. Put in the hairy melon rings and seasoning. Blanch over medium heat for about 5 minutes, or until it is 60% soft. Drain well. Soak in ice water until cool. Set aside.
3. Cut the egg tofu into round pieces, with the same thickness as the hariy melon rings. Stuff the egg tofu into the rings. Put on a round plate. Keep in a refrigerator.

On the day of feast

1. Cut the broccoli into small floral. Rinse well.
2. Heat 1/2 cup of the vegetarian stock (refer to p.210 for method). Pour into the plate containing the hairy melon rings. Steam with a lid on for about 10 minutes.
3. Pour 2 cups of hot water into a small pot. Add the seasoning for broccoli. Put in the broccoli and blanch for about 4 minutes. Dish up.
4. Take out the hairy melon rings. Pour the vegetarian stock (from hairy melon rings) into a wok. Mix in the seasoning for thickening glaze. Pour on the hairy melon rings. Put the broccoli on the side of the plate. Serve.

211

Beancurd Skin Rolls with Vegetables

Tips for shopping

- Buy thin, round and translucent beancurd skin of about 2 1/2 feet in diameter. It is generally available at soy food shops at about $4 to $5 each. Beancurd skin from Shu Kee in Sham Shui Po is high quality.
- Beancurd skin should be kept refrigerated in a zip lock bag to avoid drying and being rotten.
- Use hairy wood ear fungus, which is crunchier.

Ingredients

2 big sheets beancurd skin
10 dried black mushrooms
10 button mushrooms
150 g straw mushrooms
10 skinned water chestnuts
120 g celery
100 g carrot
1 small bundle mung bean vermicelli
1 ear wood ear fungus
* dried black mushrooms, straw mushrooms and water chestnuts are taken from vegetarian stock, pls refer to p.210.

Seasoning

1/4 cup water
1 tsp salt
1/2 tsp vegetarian chicken bouillon powder
1 tbsp vegetarian oyster sauce
1 tsp sesame oil

Thickening glaze

3 tsps caltrop starch
3 tbsps water
* mixed well

One day before feast

Method for vegetable stuffing:

1. Soak the mung bean vermicelli in water to soften. Cut into small sections. Drain.
2. Cut the black mushrooms, skinned celery, carrot and wood ear fungus into fine strips.
3. Finely slice the button mushrooms, straw mushrooms and skinned water chestnuts.
4. Put all the vegetable stuffing in hot water. Boil for about 3 minutes and drain.
5. Heat a wok. Add 2 tsps of oil. Roughly stir-fry the vegetable stuffing. Add the seasoning. Give a good stir-fry. Gently mix in the thickening glaze. Let it cool down. Keep refrigerated.

On the day of feast

1. Cut the whole sheet of beancurd skin into 10 small rectangles (about 4.5 inches x 6 inches). Put into a bag to avoid drying.
2. Mix 1/2 egg wash and 2 tbsps of caltrop starch into egg batter. Set aside.
3. Lay 2 tbsps of the vegetable stuffing on the small sheet of beancurd skin. Wrap and roll into a bar shape (7 cm x 3 cm). Brush the egg batter on the edge to seal the roll.
4. Heat a wok over medium heat. Run some oil on the wok. Remove the oil. Adjust to low heat. Fry the beancurd skin rolls until both sides are golden and crisp. Dish up. Sop up the oil. Serve with Worcestershire sauce or Zhenjiang vinegar.

Stir-fried Lily Bulbs and Corn Kernels with Vegetarian Scallop

Tips for shopping

- Fresh lily bulbs are generally vacuum-packed. About 2 packs are needed for this recipe. It is ideal for picking those without brown and withered surface.
- Buy frozen corn kernels or mixed vegetables (corn kernels, diced carrot and peas) at frozen food shops or supermarkets. Or buy fresh whole corn and take the kernels after steamed.

Ingredients

2 bundles Enokitake mushrooms
150 g fresh lily bulbs
150 g celery
150 g corn kernels
120 g carrot

Seasoning

1/2 cup water
1/2 tsp salt
1/2 tsp vegetarian chicken bouillon powder
1/2 tsp sugar
1 tsp oil

Thickening glaze

1/2 tsp salt
1/3 tsp vegetarian chicken bouillon powder
1 tsp vegetarian oyster sauce
1 tsp caltrop starch
3 tsps water

213

One day before feast

- Tear the petals off the lily bulbs and rinse. Shave off the withered part with a small knife. Keep refrigerated.
- If the whole corn is used, steam to make it done and then take the kernels off. Keep refrigerated.

On the day of feast

1. Cut away the end of the Enokitake mushrooms. Cut into small sections of around 1/2 inch long. Loosen the mushrooms. Rinse and drain.
2. Shave the skin off the celery and rinse. Dice the celery and carrot together (about 1/2-inch dices). Set aside.
3. Heat a wok. Put in some oil. Deep-fry the Enokitake mushrooms over medium heat until golden. Drain.
4. Put 1/2 cup of water in the wok. Add the seasoning. Bring to the boil over high heat. Put in all the vegetables and stir-fry for about 2 minutes. Set aside.
5. Heat a wok. Put in 2 tsps of oil. Stir-fry the vegetables evenly. Add the mixed thickening glaze. Give a good stir-fry. Put on the plate. Sprinkle the deep-fried Enokitake mushrooms (deep-fried scallops) on top. Serve.

Pumpkin Bowl with Coconut Milk

Tips for shopping

- Select deep orange colour and round shaped Japanese pumpkin with heavy weight. It should have tough skin and a dry stalk (showing it is ripe and sweet). You may buy early as it can be kept for 1 to 2 months in a cool place.
- Use fresh button mushrooms instead of canned ones. It is fresher in taste.
- Vegetarian shrimp meat is available at vegetarian food shops. It is majorly vacuum-packed.
- Better select taro with a dry body to avoid buying watery taro.
- Canned coconut milk can be used.

Ingredients

1 Japanese pumpkin
150 g fresh straw mushrooms
10 button mushrooms
10 skinned water chestnuts
100 g carrot
1 pack vegetarian shrimp meat
300 g taro
2/3 cup vegetarian stock (refer to p.210)
1/5 cup coconut milk
* button mushrooms and water chestnuts are taken from vegetarian stock, pls refer to p.210.

Seasoning

1/2 tsp salt
1 tsp vegetarian chicken bouillon powder
1/2 tsp sugar

One day before feast

Method for pumpkin bowl:

1. Rinse the pumpkin. Cut across the top in about 1 inch thick as the pumpkin lid. Remove the seeds and pith. Rinse the inside.
2. Steam in a wok for about 8 minutes, or until it is 50% soft. Take out. When it cools, coat with cling wrap. Keep refrigerated.

On the day of feast

1. Take out the pumpkin bowl from the refrigerator. Rest in room temperature.
2. Cut the button mushrooms and water chestnuts in half.
3. Peel the taro. Cut into 2 cm dices. Coarsely dice the carrot.
4. Put 2 cups of water in a wok. Bring to the boil. Add the straw mushrooms, button mushrooms, carrot, water chestnuts, vegetarian shrimp meat, 1 tsp of salt and 2 tsps of ginger juice. Boil for about 3 minutes. Drain and set aside.
5. Heat the wok. Put in about 2 tsps of oil. Fry the taro over medium heat (about 2 minutes). Pour in 2/3 cup of the vegetarian stock. Add all the ingredients. Bring to the boil. Turn to low heat. Simmer until the surface of taro starts to dissolve. Put in the seasoning and boil for 3 minutes. Mix in the coconut milk. Transfer to the pumpkin bowl.
6. Steam the pumpkin bowl for about 12 minutes. Serve.

Stir-fried Two-colour Rice with Diced Vegetables

Tips for shopping

- Vegetarian luncheon meat can be bought at vegetarian food shops. Majorly vacuum-packed, it is offered in slices or whole. Buy whole meat and cut into dices.
- Raw pine nuts are available in bulk. Or buy vacuum-packed roast pine nuts in supermarkets to save work.

Ingredients of steamed rice

2 1/2 cups white rice*
1/2 cup red rice*
3 cups water
* measured with the cup that comes with the rice cooker.

Ingredients of fried rice

200 g vegetarian luncheon meat
100 g flowering Chinese cabbage
100 g carrot
5 black mushrooms (taken from vegetarian stock; refer to p.210)
50 g pine nuts
2 eggs

Seasoning

1 tsp salt
1 tsp vegetarian chicken bouillon powder
2 tsps light soy sauce

One day before feast

Method for steamed rice:

1. Wash the red rice. Soak in 1 cup of water for about 3 hours.
2. Wash the white rice. Put into a rice cooker with the red rice (with soaked rice water). Add 2 cups of water. Cook until done. When it cools, keep refrigerated.

Method for pine nuts:

Heat a wok. Stir-fry the pine nuts without oil over low heat until the surface is light golden (about 4 to 5 minutes). Dish up. Let it cool down. Store the pine nuts in an airtight bottle or box to keep away from moisture.

On the day of feast

1. Take out the steamed rice from the refrigerator in advance, or heat up in a microwave oven.
2. Dice all the ingredients and rinse. Blanch until done. Set aside.
3. Whisk the eggs. Heat a wok. Put in 2 tsps of oil. Turn to medium heat. Stir in the egg wash until half done.
4. Put in the steamed rice and swiftly stir-fry. Turn to low-medium heat. Stir-fry until the rice is hot. Add the ingredients and stir-fry until hot. Put in the seasoning and stir-fry for about 1 minute. Finally sprinkle with the pine nuts. Give a good stir-fry. Serve.

215

Sweet Osmanthus Pudding with American Ginseng

Tips for shopping

- Sweetened osmanthus contains osmanthus flowers soaked in syrup. It can be bought at groceries, commonly found in Chaozhou food shops or groceries in Kowloon City market. Generally, both are contained in a small glass bottle costing at around $20 to $25.
- Sliced American ginseng and Qi Zi are available at Chinese herbal medicine shops.

Ingredients

5 g sliced American ginseng
20 g Qi Zi
1 tsp sweetened osmanthus
3 tbsps osmanthus syrup
(from sweetened osmanthus)
60 g brown sugar
2 cups water
20 g gelatin powder

One day before feast

Method for sweet osmanthus pudding with American ginseng:

1. Soak the gelatin powder in 100 ml of water for about 5 minutes, or until the gelatin powder swells up by absorbing water. Set aside.
2. Soak Qi Zi in water until soft. Drain.
3. Bring 2 cups of water to the boil. Turn off heat. Add the American ginseng and soak for about half an hour with a lid on.
4. Heat the American ginseng water. Add the osmanthus syrup, sweetened osmanthus, brown sugar and Qi Zi. Mix well.
5. Put the gelatin powder in a microwave oven. Turn to medium heat. Heat up for 10 seconds. Put into the above sweet soup. Mix well. Pour into a cake mold or deep dish. When it cools, chill to set.

On the day of feast

Take out the sweet osmanthus pudding from the refrigerator. Cut into pieces. Serve.